D0225517

Evolutionary Systems and Society

Evolutionary Systems and Society

A general theory of life, mind, and culture

Vilmos Csányi

A publication of the General Evolution Research Group

Duke University Press Durham and London 1989

Printed in the United States of America
on acid-free paper ∞
Library of Congress Cataloging-in-Publication Data
appear on the last printed page of this book.

I am weary of your quarrels,
Weary of your wars and bloodshed,
Weary of your prayers for vengeance,
Of your wranglings and dissensions;
All your strength is in your union,
All your danger is in discord;
Therefore be at peace henceforward,
And as brothers live together.
—Henry Wadsworth Longfellow

Contents

Foreword

I first met Vilmos Csányi at one of the early "Fuschi Conversations" of the international systems community. Much impressed by his contributions to our discussions, I began to study his research findings on evolutionary theory and read his book on a General Theory of Evolution. The more I learned about his scholarship, the more convinced I became that what he had accomplished in his many years of contributions to evolutionary science should be brought to the attention of the broadest possible audience. I suggested to Vilmos that he should develop a comprehensive exposition of his research findings for publication. He accepted my suggestions, completed his new manuscript, and Duke University Press is now publishing his work.

As a researcher in systems science and societal evolution and as a practitioner in the application of systems ideas and evolutionary theory in societal systems, I have found this book well grounded in systems theory. It offers new insights, much novelty, and exceptional power of reasoning in explaining general evolutionary theory as well as societal evolution.

What is of particular significance to the reader is the hopeful vision that Professor Csányi is projecting for the purposeful evolution of the global human society. I am convinced that readers will not only be enlightened by the ideas presented in this work but also inspired by the vision set forth in the book.

Béla H. Bánáthy

Acknowledgments

I am very grateful to a number of friends who encouraged me to pursue this subject through several years, especially John Szentágothai and George Marx. I also am thankful to György Kampis who helped me clarify several issues during our common work. I benefited greatly from discussions with András Szöllösy-Sebestyén, and also John B. Corliss, who helped with the revision of the English text. I would like to thank Antal Dóka for the nice drawings that appear throughout the book and Judit Gervai for helping me scrutinize style and readability.

Finally, I pay special tribute to Béla Bánáthy whose deep interest and encouragement made this work possible.

Preface

I have been studying evolutionary systems since the middle of the 1970s. Results of this work were published first in Hungarian (Csányi 1978, 1979) and later in English (Csányi 1980, 1981, 1982) as a general theory of evolution. I sought to formulate the mechanisms and laws of biological evolution within a general framework so as to search out and characterize analogies at the higher levels of biosocial organization.

The laws of a general evolutionary process inherent in Nature were first formulated by Spencer (1862), but a favorable intellectual and spiritual environment for such a theory, encompassing both natural and social sciences, seemed to develop only in the past few years.

With regard to its main tenets this second version of my general theory of evolution is identical with the first, but it has been extended to include new results, which were born during the refinement of some of its concepts and our further, recently published theoretical work (Csányi 1985, 1987a, 1987b; Csányi and Kampis 1985; Kampis and Csányi 1985, 1987a, 1987b, 1987c). The most important development is that I have attempted to fit my study of evolutionary processes into the framework of the General Systems Theory developed by Bertalanffy and others (Bertalanffy 1968). This, I believe, makes a more exact and rigorous formulation possible.

In the first chapter I deal with some system-theoretical problems that are necessary for the study of evolution, such as the definition of system, model, organization, information, etc., without attempting a comprehensive treatment. I also introduce the model of autogenesis that serves as a tool to describe real evolutionary systems. In later chapters the origin of life and molecular, organismic, ecological, and sociocultural evolution are treated. In the last chapter I try to survey some possible practical uses of the theory.

There are three appendixes that serve to clarify some important general questions regarding evolutionary theory. Appendix 3 is the work of my friend and coworker, György Kampis.

1 Theoretical Problems of Modeling the Biosocial System

System and model

The only undisputed concept in Systems Science is probably that the Universe is the totality of reality. This concept has two equivocal meanings, namely, "that which is out there" and the concept itself created by our minds. "Out there" can be dealt with only through our conceptual means, and this dichotomy cannot be argued away by any sophisticated philosophy. Therefore, the definitions that follow serve only as practical tools.

Real beings "out there" are called *natural entities* if criteria can be found to distinguish them from other entities, which are frequently called their *environment*. An entity can be a star, a mountain, an atom, an organism, or many other things. Nevertheless, the classification of an entity, more or less independently from our criteria, is always a subjective process. An entity can be called a *system* if it embodies a finite set of subentities, named *components*, and if there are *interconnections* among these components by which it is possible to distinguish this entity from others, i.e., from the environment. System and component mutually determine each other. The component as an individuum is determined by its interactions and does not exist per se. The component's individual existence is achieved by the existence of the system; nevertheless, the system itself emerges by the sum of the interactions of its own components.

It follows from the above that a natural entity is called a *natural system* if it can be distinguished from its environment by criteria based on the interconnections of its components. Of course, this definition does not exclude interconnections between a natural system and its environment. Examples of natural systems are volcanoes, rivers, cells, organisms, ecosystems, and human societies.

Human (and, in general, animal) knowledge-gaining processes always are based on the use of systems *models* (Csányi 1985c). A model by systems-theoretical definition is always a *simpler* system that simulates a more complex system in certain aspects of its behavior (Mesarovic 1964). The model's components and their interconnections reflect the components and interactions of the complex system. Therefore, the process of model construction is always a kind of identification between the complex system and its model in which simplification is inherently unavoidable. (A model is used by being operated and accordingly leads us to predict the behavior of the modeled complex system.)

The framework of thinking in the natural sciences is based on the use of such models. Reality is the totality of enormously complex, mutually interconnected processes. Scientific research attempts to separate and isolate phenomena and effects. It searches for the simplest connections that can be used to describe and represent the phenomenon under investigation. Science makes simple models that reflect only a part of reality but that make thinking about reality possible. A model fulfills its role if, even only in a humble way, it is suitable for *practical predictions*.

Newton in his famous "laws" did not explain the nature of physical forces; he only made a dynamic model of them, and based on this model he predicted some properties of moving bodies. It is well known that these predictions are more or less valid in a large domain of reality with well-recognized boundaries, yet we still do not *know* what a physical force is. Newton's laws are not explanations in the absolute sense of the word; they are only models suitable for predictions within limits. If we turn to the history of natural sciences, we see that the specific constructions of scientific models are always occasional. A model's value is judged by its practical use. There are no philosophical or other criteria that help us select a model before its application (Lakatos 1976).

The more complex a natural system, the more difficult it is to construct a model of high predictive capability. Usually, mathematical models are most appreciated because of their unambiguous logic, relative structural simplicity and compactness, and well-developed deductive verifiability. Nevertheless, these elegant tools have some inherent drawbacks. They are not flexible enough, are often not suitable for describing truly complex phenomena, and are not creative in a broad sense. Therefore, the first step of model-building should always be the construction of an intuitive model, which by its nature is able to embody ambiguities, contradictions, and descriptions on different levels. An intuitive model thus may reflect a sufficiently complex part of reality and at the same time allow the possibility of a further, more rigorous mathematical treatment.

Theories of enormous influence, such as the theory of atoms in physics or Darwin's theory of evolution, were nonmathematical, and their impact has far exceeded those occasional mathematical constructions that appeared later and covered only particular aspects of the theories. It also is well known that logico-mathematical theories are tautological in the sense that they are derived analytically from a set of axioms, and therefore they are unsuitable for proving the validity of an intuitive theory. In many cases the bright armor of mathematics only helps to delay the recognition of inherent weaknesses in an intuitive theory. This does not mean that mathematical models are useless in science, particularly in biology; I wish only to emphasize the basic primacy of intuitive models relative to mathematical models, which are only auxiliary tools not able to surpass the former. The takeover of quantum mechanics in physics was not due to the mathematical models developed inside the realm of Newtonian physics but to intuitive ideas.

The biosphere of Earth, including human society, can be viewed as a single natural entity. If we accept this as a working hypothesis, we may attempt to examine whether this entity is a system, and to what extent. To answer this question we have to use concepts such as identity, organization, component, function, etc. Most of these terms have been defined and discussed in previous research and the *autogenetic model* of the biosocial system also has been described (Csányi 1985; Csányi and Kampis 1985). In these articles comparisons were made between the autogenetic model and the theory of autopoiesis (Maturana and Varela 1980; Varela et al. 1974) and also with the theory of the process of life (Haukioja 1982). Here I will not discuss the problems of comparisons again, but concepts essential to the autogenetic model are given.

System, components, identity

The first question is whether the biosocial entity (living beings together with human societies) can be considered as a *unity*, and whether or not this unity is a system.

Let us begin with the simplest facts. Components of the biosocial entity are organisms, i.e., humans, plants, animals, microbes, etc., and the artifacts made by man. What makes these things components is their almost innumerable interactions. Many branches of biology deal with the interactions among organisms, so we do not have to go into detail here. Humans are organisms, too; they are unable to live without interactions with other biological components (feeding, digestion, breathing, etc.).

It also is evident that humans and their artifacts are continuously interacting. Moreover, interactions among man-made artifacts and nonhuman organisms are increasingly revealed in the development of agriculture and the understanding of pollution and processes of environmental protection. Large energy and material cycles that interconnect human societies with the whole biosphere can be detected; therefore, the *organizational unity* of the biosocial system also can be recognized.

The biosocial entity can be regarded as a system if we base this classification on the interactions of components as a minimal requirement of a system. The concepts of components and system are mutually presupposed. We also can assume that the biosocial system is *individual* because its network of components does not seem to be part of another similar entity.

The question of the identity and boundaries of this system is more problematic. If we regard the interactions of components as a minimal system criterion, it is easy to see that the system is underdefined. Are the rocks of the planet's surface components? A great many of them are not interacting with the other components named so far. We tend to regard them as *environment* (entities separable from the system). But if we think of the small pebbles swallowed by various birds to help their digestion of hard seeds, then we have to regard *these* pebbles as components. Similarly, limestone originates from the cytoskeleton of unicellular organisms that lived in an earlier period, and the limestone on the Earth's surface can readily join the material cycles of the biosphere; therefore, it is best to consider limestone as a component. On the other hand, the metals used by man originate from ores. Are the ores simply part of the environment, or do they become artifacts (components) during mining? And if they do become components, at what point does this occur?

I do not think it is possible to construct a model that is capable of fulfilling all rigorous system criteria because not only the system's physical but temporal boundaries are problematic. If we consider time too short—some minutes or years, for example—then many interactions may remain undetected. If we take a period too long —millions or billions of years—then we must consider enormous changes of the system's component composition, and these changes may cause the alteration of the system's *identity*.

If we define genealogical descent from generation to generation as *interaction* (see Ghiselin 1981 for clever ideas about this question), then the 3.5 billion-year-old biosphere can be considered a single entity, i.e., a single unity or system.

Identity can be defined by various criteria. We may consider the identity of a system according to the nature of its components and their interactions within a given time, but we also may include descent as a criterium, and then we might ignore qualitative changes of the components. In studying the *ontogenesis* of organisms we usually apply this latter criterium of identity. We regard the zygote, the embryo, and the full-fledged adult as a single *identical* organism, and we consider the obvious changes in the composition of its components as a preprogrammed development of the system.

Defining the boundary of the system also is not at all trivial, especially if we want to define organizational boundaries instead of topological ones. A cell is a system in itself, but it may belong to an organism that is also a system but belonging to an ecosystem, etc.

From a practical point of view it is not vital to have a system definition that fulfills all possible criteria. I find it more reasonable to accept that any system definition is a conceptual construction that can be evaluated by its practical uses. It is quite possible that the concepts of the "biosphere" or the "biosocial system" are not exact enough, but everyone knows their meaning. Quite frequently we give a system definition based on characteristics that are considered important only to find out later that it is unsuitable for studying certain problems. Then, of course, new definitions have to be created. To give an a priori final definition is theoretically impossible. If we consider only topological boundaries, we may face fewer problems. Let us consider, as a system, the physical space defined by the spherical shell (and the matter it includes) extending to a hundred kilometers below the surface of Earth and a thousand kilometers above it. All components of the biosocial system can be found in this physical space, and also many entities—atoms, molecules, etc.—that do not participate in any interconnections of importance for us. This is the price we have to pay for a topological definition instead of an organizational one. This topological definition is exact enough, taking into account the physical space and the elementary building units of the components (atoms), but it misses the organizational criteria of the system. I think it is better to use a loose, intuitive system definition if it is nice and simple and suitable for some kind of prediction.

I will not deal here with the classification of systems because in the next chapters I want to study only the biosocial system.

Structure, organization, organizational levels

It is characteristic of the biosocial system's structures that in the course of time they are *assembled* and *decomposed*. The processes of their assembly or production and decomposition represent the *network of their interactions*, which is the *organization* of the system.

Structure is the actual space-time relationship of the system's components. The same organization may appear in different structures, e.g., different physical configurations and different physical manifestations. Within certain limits the quantitative changes of the components will be considered as a structural change that does not influence the system's organization. For example, changes in the number of molecules of a given protein with a special function in a cell (e.g., enzyme induction) are structural changes that do not affect the cell's organization. On the other hand, it is an organizational feature that the cell is capable of producing this particular protein at all.

The major focus of this book is the assumption that *replication* is the type of organization that characterizes the biosocial system. Looking at the system's various components such as cells, organisms, artifacts, etc., it can be shown that these entities exist by *self-maintenance*, that is, by the continuous renewal of their own subcomponents.

The cell, for example, can be considered as a *functionally closed* network of molecule-producing processes that continually produces the same network of components and processes. The same is true for organisms or ecosystems. These entities also have a special kind of self-maintaining or "self-producing" ability, the essence of which is the continuous renewal of their components. Beyond self-maintenance is *reproduction*, the other characteristic trait of these entities. Production of the same subcomponents by a functionally closed network of processes is also the basic mechanism of reproduction. Therefore, as far as organization is concerned, self-production and reproduction belong in the same category. Furthermore, by analyzing self-production and reproduction, it can be shown that the principal mechanism in both cases is copying, that is, *replication*.

Replication is generally considered a synonym for copying, where a constructor produces a copy (replica) of a component or a given system. To do so, the constructor needs a description as the information necessary for this copying process. In the case of self-replication the total information of the constructor itself is necessary in some form. The essence of replication is functional operation, regardless of the particular mechanisms of information storage and retrieval. It does not matter whether this information is stored separately (as in the case of DNA) or is distributed throughout the entire system. It also does not matter whether a separate object (component) is copied or the constructor itself; the fact of copying lies in the functional organization, not in the particular mechanisms. Based on this argument about self-production and reproduction, the concept of replication will be used in two senses.

We define *temporal replication* as the continuous renewal of the system in time. This is the uninterrupted existence of the system, which is manifested by means of the sequential and functional renewal of the system components. The components are assembled and decomposed, but the extent to which assembly and decomposition take place is always sufficient to maintain the unity and identity of the system and its organization.

The definition of *spatial replication* is identical to that of reproduction. The system produces its own replica, which becomes separated from it in space. From one system or unit two units are formed. Spatial replication also proceeds by copying the system's components as in the case of replication in time, but in double numbers, while the original organization of the system remains unchanged.

The biosocial system is characterized by the fact that its functionally more or less closed replicative network embodies various replicative subnetworks; its components embody various subcomponents. Thus *organizational levels* are created. An organism, for example, is a more or less autonomous network of components and component-producing processes that are replicating in time, i.e., the organism is a *component system*. As a unity it also is a component of higher organizational levels, such as ecosystems or the whole biosphere. The components of the organism are the cells, which themselves are autonomous replicative networks of molecules—components of a lower level of organization. Molecules are regarded as the components of the lowest level of organization, and atoms are the *elementary building units* of the biosocial system.

Exact definition of the higher organizational levels is more difficult and will be treated in detail later. It is only mentioned in advance here that on Earth the biosocial complex is regarded as the highest level of organization and that the biosocial system is regarded as an open physical system existing in a continuous energy flow. The laws of the thermodynamics of open systems are valid, and their effect on the organization of this system is readily recognizable.

Information, function

Previously both biology and the social sciences have been dominated by open or concealed vitalistic ideas, maintaining that there were particular laws governing biological or societal processes and that these laws were different in principle from the laws of physics and chemistry. Because of the many apparent weaknesses of vitalistic ideas, after a long and bitter fight a reductionist view has gained supremacy, at least in biology. According to this reductionist philosophy, all phenomena of biological or social systems can be explained by unified principles and ultimately reduced to chemical mechanisms in spite of their obvious hierarchical organization (Crick 1967). The spectacular boom of molecular biology was regarded as proof of the reductionist view.

Without diminishing the achievements of molecular biology, this proof can be questioned. Recently disputed problems of the Darwinian theory of evolution are rooted exactly in the problem of reduction. The simple explanation of the origin of species by nucleotide changes and selection has proved insufficient. The reductionist standpoint is more and more criticized. Besides mutation and selection at the individual level, the random origin of species and particular mechanisms of species selection are assumed (Gould 1980), which cannot be reduced to a molecular level. This agrees with Stanley (1975), who emphasized that decoupling the process of microevolution (changes of the genetic architecture of populations) from macroevolution (origin of species) is unavoidable. Reductionism also has failed in the behavioral sciences in that the interpretation of acquired information (learning, culture) also seems to be bound to a higher level of organization (Plotkin and Odling-Smee 1981).

The antireductionist viewpoint was articulated most clearly by Polányi (1968), who wrote that "Mechanisms, whether man-made or morphological, are boundary conditions harnessing the laws of inanimate nature, being themselves irreducible to those laws. The pattern of organic bases in DNA, which functions as a genetic code, is a boundary condition irreducible to physics and chemistry. Further controlling principles of life may be presented as a hierarchy of boundary conditions extending, in the case of man, to consciousness and responsibility." This line of thought was elaborated in detail, incorporating newer results, by others (Pattee 1965, Rosen 1977, Primas 1977, Kampis 1986a, b, c).

In my opinion, the interrelationships of components within the various organizational levels, which we may call *algorithms*, are specific only at a given level. In a general model, in which all organizational levels are embodied, all effects computed by the algorithms of the lower levels lead to *random* events in the event-space of the higher levels. The algorithms of the higher levels represent specific nonlogical *constraints* at the lower levels. This is clearly not a vitalistic standpoint, but from it follows the conclusion that models of the biological and societal systems are different in principle from models constructed for understanding the physical world.

A simple example illustrates what I have just said. A combustion engine can use part of the free energy produced by the chemical reactions of the process for a given task determined by its designer. Combustion enhances the kinetic energy of the gas mixture's molecules, and this kinetic energy is used by the machine. We know the nature of molecular kinetic energy, and with the equations of molecular kinetics we can, in principle, predict events in this molecular space. Nevertheless, we cannot use these equations to predict the convertible useful portion of the total free energy of combustion, aside from the thermodynamic boundary as a maximal limit. We cannot use these equations or any other physical laws for designing a final perfect engine.

The engine's structure, the shape and size of its parts and their arrangement, all are *special boundary conditions* that represent *nonholonomic constraints* on the level of molecular kinetics (Pattee 1977). Therefore, the combustion engine's design is an empirical science. Various designs are tried, and those that work best are used. We cannot calculate the perfect design from the kinetic equations. An engine's structure is in principle the *description* of its special constraints.

The space of the combustion reactions is bounded with metal walls, of which one (the piston) is movable and serves to transmit power. But the description of this structure is not based on the kinetics of the molecules; it is *not reducible* to kinetics. The combustion engine must be described on two different levels. The lower level includes the kinetic equations of gas molecules, while the upper level includes the description of the specific constraints that originate from the engine's design. The working of the engine can be understood only by having both descriptions.

It is the same with a cell or a higher organism. At the molecular level these systems can be described by physical and chemical equations. But these equations do not show the constraints existing on higher levels; they do not show which *structures* constitute the boundary conditions for those physical and chemical laws.

This, of course, is not a denial of the causality principle—that the phenomena of the higher levels can be traced down to the lower levels as a *causal chain of events*. But an explanation based merely on causation is not comprehensive because it fails to describe the specific constraints existing on the higher level.

Here is another example. I look around in a shop and see an interesting article, which I pick up and examine. During this brief sequence of actions my organism goes through an enormously complex chain of events: processes of the nervous system, muscle contraction, movement of body parts, etc. Chemical and physical processes behind these events can be exactly understood and described, at least in principle. Chemical processes of the retinal membrane of my eyes, chemical processes of the nerve cells in my brain, etc., can all be described, and the causal chain behind my action can be traced. Nevertheless, this causal chain cannot explain many other features of these events. For example, why did that article catch my attention? Chemical and physical laws will not help find an answer to this very important question. But to answer it is very easy: all you have to do is ask me. The answer, of course, will not use the language of chemistry or physics; but it will use the special language of a higher organizational level—in this case the language of the psyche. Constraints at the psychical level cannot be reconstructed on the basis of description of a lower level.

The reductionist standpoint can be blamed most for its inability to use the concept of *function*, which is the most important concept in biology. The common and primary meaning of function is role, effect, transformation rule. In the component system the function is the effect or role of the action of components on the system's next higher level. Function can be determined from "above" as *constraints* exerted by processes at the higher level in the event-space of the lower level. The nucleotide triplets of DNA, for example, have a function: they direct amino acids into protein structures. A given nucleotide sequence responsible for a protein structure is not a chemically particular set at all. The DNA sequence is not determined by the laws of the chemical affinity but by the *function* of the encoded protein. Chemical affinity is only harnessed by this function in the same sense as Polányi formulated for the boundary conditions of the organizational levels. The appearance of function is always the result of a process that decouples the interrelationships embodied in the function from the properties and relations of components at the lower level. The interpretation of function becomes possible only at the system's higher levels. The function always must be formulated as an embodiment of the *description of constraints* (in the form of algorithms) in the event-space of the lower levels. Therefore, applying the concept of function in systems models needs a proper *informational theory*.

Many authors have attempted to introduce such considerations into models of biological systems. Pattee (1967) centered his argument on "traits" that were the equivalents of functions and stated that the hereditary propagation of a trait involves a description or code and therefore must involve a classification process and not simply the operation of inexorable physical laws of motion on a set of initial conditions. Rosen (1973) very clearly expressed that a dynamic model of evolution cannot be constructed without properly incorporating the concept of function. Quastler (1964) was the first to calculate the informational content of various macromolecules at the advent of molecular biology, but his finding only promoted the use of the concept of information by others as a metaphor. Very attractive models of the living cell were independently constructed on the concept of information by Pattee (1977), Liberman (1979), and Iberall (1983). The common characteristic of these models is that the interactions among their components are regarded as a kind of grammar that emerges as a linkage that can switch or evoke changed molecular states in chemical systems. It is common in the various independently developed models that biological function appears as a specific description, a kind of information relevant only in the given system, not reducible to physical processes, only harnessing them. The totality of the biological system and its functional components can be regarded as *carriers of algorithms*. Their interactions are considered as *computing processes* prescribed by these algorithms (Stahl 1965).

The question immediately arises: what is the "meaning" of these algorithms? In this book I try to establish my hypothesis that these algorithms are the algorithms of a *self-replicating process*. A model representing biological systems must be constructed so that it represents the replication process of both the components and the whole system. The model offered is the *component-system model*, a replicative organization in which the interaction of the components leads to the maintenance of the whole system and occasionally to the replication of the system in space (reproduction).

I use a concept of information to describe the component systems that was analyzed in detail by Kampis (1986a, 1986b). Kampis distinguishes *effect* and *knowledge*. Knowledge is a *description* that has meaning only for the observer, while the effect—the final source of knowledge, according to Kampis—is the specific way a given component functions in a given system. Based on effect and knowledge, Kampis defines two kinds of information. The structural arrangement responsible for the effect of a given component and the effect itself are expressed in the concept of *referential information*. The "referential" notation points to the fact that for expression of an effect by a given component the whole system is necessary. *Nonreferential information* describes both the effect and the structure responsible for it. Generally in this book the latter concept of nonreferential information will be used.

Consequently, information necessary for the replication of a system is carried by its components, and this *replicative information* contains both structural and functional information. The *organization* is the mode of the components' interaction, and it creates the link between structure and function. Organization is closely connected to function, but we use organization when dealing with the whole system, and we use function when we describe the interactions of given components. Components carry only a part of the whole system's replicative information, collectively making up the information that allows for the whole system's replication.

In this framework the referential aspect of a given component's replicative information as represented by its structure and function forms an inseparable unity. This information is the mode of effect of the given structure, and its meaning is relevant only within the whole system; i.e., it is a system-dependent parameter.

Function can be described by an organizational approach. The function (effect) of a replicative organization's system is replication. The system is working toward its own replication. The function considered in such a way will be called *replicative function*.

The function of the components can be deduced from the behavior of the whole system in a general way and a specific way. In general, a replicative system's components are the entities that through their interactions participate in the whole system's replication. It is evident that they have replicative functions because they promote both the system's and their own replication. Therefore, the *general function* of these components is their role in the replicative process.

On the other hand, considering interactions among components, we always find special mechanisms that influence the *probability of genesis* of these participating components. A digestive enzyme in a cell, for example, splits the structure of other proteins (components); i.e., it decreases the probability of their existence. A predator taking its prey also decreases the probability of existence and genesis (of progeny) of the prey animals.

The particular mechanisms are different in various component interactions, but through interactions the components always influence the probability of one another's genesis (in one or both directions). Influencing the probability of the components' genesis also is a function—a general function closely related to the replicative function.

The problem of a system's *autonomy* also is related to the concept of information. Individual organisms always show autonomy to a certain degree; that is, they are separated from their environment and are not entirely controlled by the environment. At a higher, supra-individual level of organization the degree of autonomy may be different.

The biosocial system is characterized by complex, hierarchical interactions of functions ranging from molecules to the totality of the system. Any effect that causes a change in the functional network of the system's replicating components may be regarded as a functional effect because the same components also are *building blocks* at a higher organizational level. Therefore, an individual component's autonomy is always relative. Autonomy in the full meaning of the word exists only at the highest level of the biosocial system.

Origin of information and complexity

Models in physics usually describe simple interactions, or, more precisely, interactions that appear to be simple, and the models themselves are not very complex. Nevertheless, the predictive value of the equations describing interactions of atoms, electrons, and photons are high, as is shown by the modern technology founded on physics.

In a biological system, describing even the simplest interactions requires highly complex models operating with many elementary units and system parameters. There are thousands of chemical reactions in the simplest cell. Several million atoms can make up a macromolecule. Bodies of higher organisms consist of billions of cells. The number of various species living on Earth may well be over 30 million, according to recent estimates. These examples show the enormous variability of the biological system's components. The possible interactions of these components are more numerous by several orders of magnitude. Biological systems therefore have a special character (at least in the way we perceive them) that has to be considered if we want to construct a specific systems model. Let us call this character *complexity*.

The concept of complexity can be satisfactorily understood on an intuitive level, but it has no generally accepted definition. In our treatment we use Bunge's (1963) definition, which distinguishes *ontological* and *semiotic* complexity. Ontological complexity originates from the inherent nature of the interactions of matter, which are infinite and immeasurable, and we can approach it only by using models. Semiotic complexity is the self-complexity of the models, which were made to represent reality. This kind of complexity is finite and can be measured in the same way as specific digit sequences (Kolmogorov 1965, Chaitin 1966). From this definition it also follows that there are many kinds of semiotic complexity, e.g., structural and functional, which we will treat below (Kampis and Csányi 1987a).

There is a close connection between semiotic complexity and our concept of structural information. Description of a molecular or other structure represents measurable "semiotic" information, which we call structural information, and thus semiotic complexity can be expressed or measured by it.

The definition of complexity by Saunders and Ho (1976, 1981) is similar to that of semiotic complexity, and they proposed that it is the only continuously growing parameter of the biosphere during evolution. Wicken (1978, 1979) also has studied the growth of complexity in connection with the entropy concept of thermodynamics. He found close relationships among energetic changes, increase of entropy, and the change of complexity in physical systems. The essence of his argument is that in spontaneous processes the increase in heterogeneity is inevitably accompanied by an increase in complexity. Thus the increase in heterogeneity is equivalent to complexity.

Large heterogeneous molecules can form spontaneously from smaller, simpler ones at the expense of the system's free energy, because reactions producing complex end products increase the entropy. This line of thought is easily understood if we consider that disorder and molecular heterogeneity are the same concepts in thermodynamics. The structure of a protein molecule that is built up from a single amino acid is clearly more ordered than one built from many different kinds of amino acids. Wicken showed that the entropy of a system containing large heterogeneous molecules is greater than that of one consisting of a few simple molecules. This is why the formation of larger heterogeneous molecules is more probable. A system containing large heterogeneous molecules is more complex than a system containing simpler or homogeneous ones. Therefore, because of simple thermodynamic considerations, an open system with spontaneous chemical processes always tends to change in the direction of higher complexity. The Universe in this sense is equivocally creative.

Acceptance of this very remarkable idea was probably delayed by the fact that a more heterogeneous, more complex system is not necessarily more *organized*. Organization, as we showed, is the consequence of *specific constraints* and is not closely connected with changes in entropy. In open systems complex molecules may emerge, but their complexity is *unorganized* because of the absence of a system that would embody the complex molecules as components—more precisely, as components with functions. Wicken stated that for life to emerge, a functional organization is necessary, but he did not examine the origin of organization.

I regard complexity that originates in the molecular structure as *structural complexity*, while the complexity emerging from an existing organization will be called *organizational complexity*. The two must be clearly distinguished. The difference can be seen most clearly in the case of an organized system. The organized system (for example, a cell) consists of complex components (structures); therefore, part of its total complexity is structural, which can be computed from the structural complexity of its components. The other part of this total complexity comes from the organized network of component interactions. This functional complexity can be calculated from a description of the functional network.

Both kinds of complexity can be interpreted within the replicative model. An already existing organization, even the most primitive, is capable of transforming an unorganized system. If we randomly change the components of a replicative cycle or network, removing or introducing components, then the original replicative organization acts as a *selective agent*. Changes that do not interrupt the replicative organization may be incorporated into the system and increase its complexity without altering its organization. However, changes that damage the replicative organization replicate poorly or disappear from the system.

It will be shown in the second chapter that there are systems, built up from a few kinds of simple molecules, that are capable of replication and may organize larger, unorganized systems. Such a primitive, small system, which can organize a larger one, is called a *system precursor* after Gray (1975), who originally worked out the concept for autopoietic systems in psychiatry.

No matter how primitive the replicative system is, it is able to incorporate new components, and this always results in the emergence of *new functions*, because new components can join the replicative system only through some interaction, through a function that can fit into the already existing organization.

The more primitive the initiating system precursor, the more limited is the set of possible components and functions. Only a few components can join the replicative cycle. With growing complexity the number of potential components that can fit into the system becomes larger, and complexity will grow ever faster.

Structural information emerging in an unorganized open system will be called *parametric information*, or parametric complexity, because this kind of complexity emerges as a result of the system's general parameters such as temperature, concentration, etc., and is unorganized. If there is a system precursor present in the system, then the complexity due to the effect of this organizing mechanism is called functional complexity and its information content is *functional information*.

For example, the DNA of a cell is a complex structure, which has *structural information*, i.e., the description of its atomic arrangement, and it contains functional information that is the genetic code. The latter corresponds to a *separate description*, independent of reaction rates (Pattee 1977), which represents a prescription for the interactions among components. This separate description can be interpreted only in a given system, but it is objective and exists independently of the observer.

In the above sections we discussed the most important concepts necessary to follow our later discussion. In the following sections we introduce the autogenetic model that has been formulated on the basis of the above considerations.

The replicative model

Definitions and explanations

System

A system is a finite physical space, separated from the background by topological or organizational boundaries, in which components with physical structures and the building blocks required to create these components are present. Components are assembled and disassembled continuously, and there is an energy flux flowing through the system that is capable of exciting some of the building blocks. The number and types of building blocks present in the system—the

Figure 1 Main features of a system.

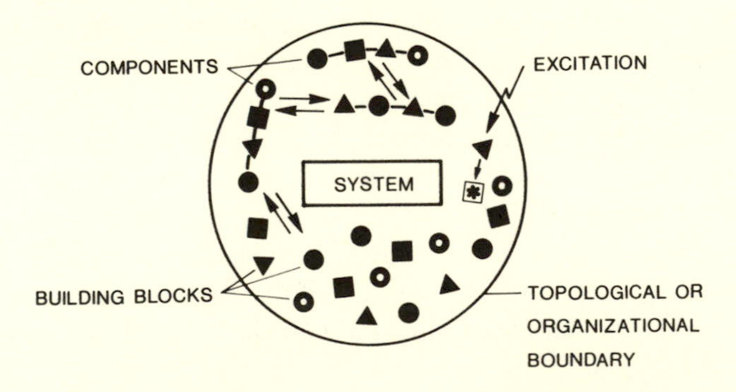

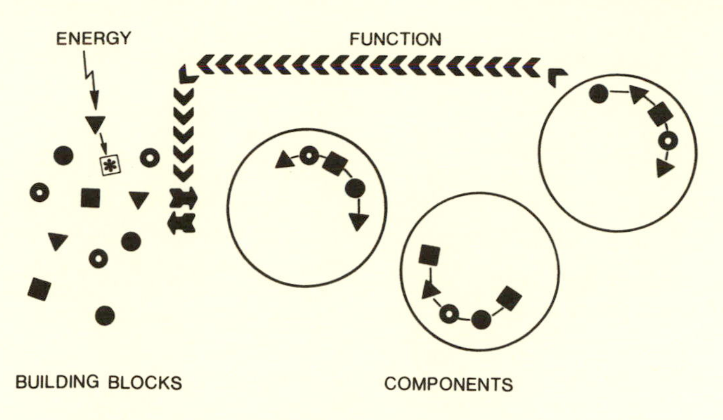

Figure 2 Assembly and disassembly of components.

quality and intensity of the energy flux, etc.—constitute the system's parameters (figure 1).

Zero-system

A system of components that have not yet developed functions is considered to be a zero-system. A zero-system has no organization.

Function

Function is the ability of components to influence the probability of genesis or survival of the system's other components due to their relationships with the component-producing or component-decaying processes (figure 2).

Information

The description of the components by the arrangement of their building blocks is *structural information*.

That part of the structural information of a system's components that is only the manifestation of the system's physical parameters is *parametric information*. It is only a nonreferential kind of information.

Description of functions and underlying interactions (mechanisms)

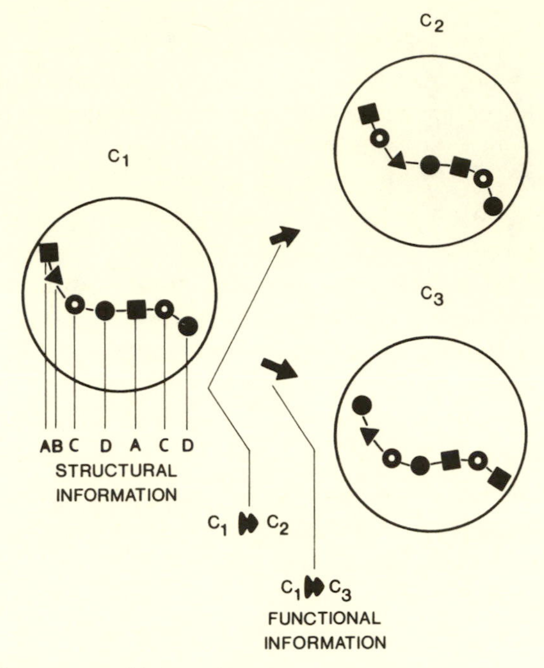

Figure 3 Different types of information.

is *functional information*. It always has both referential and nonreferential aspects (figure 3).

Replicative function, replicative information

In the biosocial system the replicative function is the most important of numerous possible functions. This term specifies a property that allows a component (or a set of components) bearing a replicative function to increase its (their) own probability of genesis or survival in the system. The description of the interactions of components that are necessary for replication is regarded as replicative information.

Organization

The hierarchical network of component interrelationships and component-producing processes, i.e., the network of functions, constitutes the organization of the system.

Replication

If the system's organization allows by various copying mechanisms the network of components and component-producing processes to produce partly or totally the same network again, this is called replicative organization. In the replicative process both the system and its components are produced. Replication is a copying process; a constructor produces a copy (replica) of a component or a given subsystem. In a replicative organization components are endowed with functions, which are expressed as functional information. In the component-producing process regenerating the system, this information also is regenerated; thus it is literally a self-copying process. Two forms of replication are distinguished. *Temporal replication* is defined as the system's continuous renewal in time by the sequential and functional renewal of the system's components while the unity and identity of the system are maintained. *Spatial replication* is identical to reproduction. The system produces its own replica, which becomes separated from it in space. From one unit two units are formed (figure 4).

Fidelity of replication

Replicative processes are never error-free. Replication can thus be characterized by its fidelity. If replication is precise by all available

Figure 4 Replication in space and time.

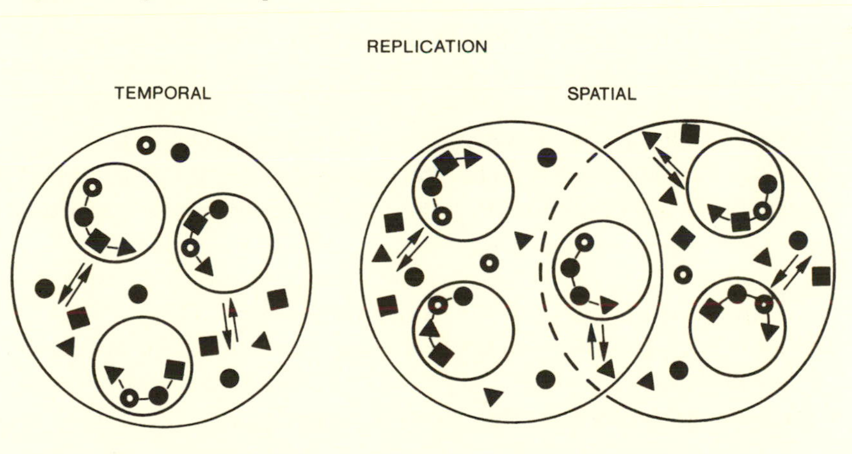

measures, it is called identical replication. If fidelity can be expressed by a number, e.g., from zero to one, replication is nonidentical. In the case of nonidentical replication either the structure of the components or the system's component composition is different from the preceding state.

Autogenetic System Precursor (AGSP)

AGSP is a minimal set of components that is able to replicate and that fulfills the following criteria: (a) it contains at least one cycle of component-producing processes; (b) at least one of the components participating in this cycle can be excited by the energy flux flowing through the system.

Operation of the replicative model

On the basis of various considerations and data collected from real biological and social systems, which will be discussed, it has been inferred that the functional informational content of a zero-system containing an appropriate AGSP will increase in time, with the simultaneous decrease of parametric information. This process has been called *autogenesis* (Csányi 1985, Csányi and Kampis 1985).

As time advances, an increasing part of functional information becomes replicative information. This can appear only as an extension of AGSP; that is, additional replicative cycles appear that are interconnected with the AGSP. Their formation is termed *supercycles*. As time passes, replicative coordination of supercycles develops and fidelity of replication increases. We can speak of some kind of functional differentiation and cooperation that result in the formation of communities of simultaneously replicating components; that is, subsystems called *compartments* form. The components of these are separated from the components of others by their participation in coreplication. The emergence of compartments is equivalent to the organizational, i.e., functional, closure of the network of component-producing processes and components having a replicative function. This succession of events is called the *compartmentalization* and *convergence* of replicative information.

Components and compartments embodying them may coexist with

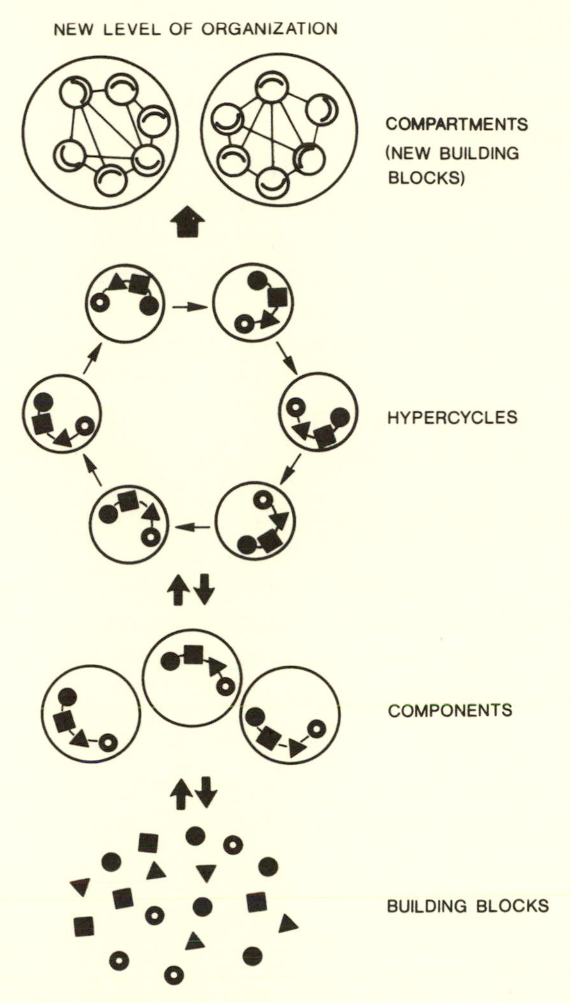

Figure 5 Formation of organizational levels.

different levels of replicative fidelity. For a while components may replicate with high accuracy, but the compartments formed by them do so with a low level of fidelity. As time passes, the fidelity of replication of both may increase, and a *next level of organization* may be created; a "compartment of compartments" is formed, which also replicates with an increasing fidelity. A system may contain several different kinds of compartments, which are all replicative units with diverse fidelity. Among these, interrelationships develop, and as a result their replication becomes coordinated. Gradually, the whole system will start replicating as a final replicative unity. In the autogenetic process the system's organization (and its parts) changes due to the functions of the emerging new components. Thus autogenesis is possible only while the state of identical replication has not yet been achieved. In that state the system is functionally closed, and its replication in time continues as long as the environment does not change. There are no further organizational changes initiated by organizational causes because new functions cannot originate. In the state of identical replication the system is an autonomous, self-maintaining unity, a network of components and component-producing processes that through the functional interaction of its components produces exactly the same network that produced it. Its organization is almost closed and cyclic. Its input and output are subordinated to its replication, but through them its existence depends on the environment's invariance. The notion of function stresses that the end-states of autogenetic systems are not simply the fixed points of some dynamic processes. Autogenesis is the evolution of active self-construction (figure 5).

2 Molecular Evolution

The origin of life

Since the first serious theory on the origin of life was born (Oparin 1924), the study of evolution has become a broad and flourishing field. Geochemical research has shown that biological evolution on Earth started 3.8 to 3.4 billion years ago. Traces of advanced chemical evolution have been successfully identified in 3.8-billion-year-old fossils (Pflug and Jaeschke-Boyer 1979, Schuster 1984). We have evidence of the existence of primitive organisms 3.5 billion years ago (Walsh and Lowe 1985). Traces of biochemical activity of prokaryotes with modern enzyme apparatus have been detected in rocks 2.3 billion years old (Oehler et al. 1972, Nagy 1976). The basic problem of scientists engaged in research on life's origin is the theoretical and experimental reproduction of events in the course of which complex molecular systems capable of self-reproduction evolved from simple carbon compounds. Early theories generally avoided this question out of the conviction that in the course of a single, incidental event a self-reproducing system had spontaneously "arisen," and this is how evolution began (Beadle and Beadle 1966, Monod 1971). Others avoided the question by placing the origin of life in space outside Earth (Crick and Orgel 1973). Statistical methods can prove that the possibility of the accidental appearance of self-reproducing structures consisting of linear polymers is extremely small, so that this explanation cannot be accepted as a scientific hypothesis (Wigner 1961, Morowitz 1968, Argyll 1977). The view is spreading that the evolution of life is a necessity, and the physical and chemical processes leading to its emergence can be experimentally reproduced (Darlington 1972, Kuhn 1976, Olson 1981, Matsuno 1984).

Synthesis of building blocks: polymers on primeval Earth and in outer space

For a long time students of the origin of life have searched for those conditions that permit the abiogenic synthesis of the basic biomolecules—the "building blocks" of life—such as sugars, amino acids, purines, and pyrimidines. The physical and chemical parameters for synthesizing these substances were found, and each group of compounds could be prepared under laboratory conditions (Fox 1965).

The questions of the origin of life and the origin of the building blocks have recently received a new twist, and it is quite possible that we will have to reconsider all our theories about the relationship of life and the Universe. Earlier it was shown that traces of the first steps of biochemical evolution could be demonstrated in cosmic space (Robinson 1976). New data and theories show that processes connected with the origin of life also have a major role in the origin of the stars. We do not yet have all the evidence, but the little evidence we do have seems to support an entirely new picture of the evolution of life and the Universe.

The Universe was probably born in a violent explosion (the "Big Bang") some 15 billion years ago. After early, extremely fast evolution (approximately 0.7 million years) two elements, hydrogen and helium, made up most of the Universe's mass. All other elements are by-products of later stellar evolution. At the end of its first period the Universe was an enormous cloud of hydrogen (75 percent) and helium (25 percent) extremely hot at the start. Then it began to expand and cool, creating favorable conditions for local condensation processes, and thus the Protogalaxies came into being. The Protogalaxies began to contract under their own gravity, giving birth to the short-lived giant stars, each having a mass larger than a hundred suns. In these giant stars helium and other heavier elements were formed from hydrogen, and after only a million years they exploded in violent supernova bursts. The material produced by them was scattered throughout the Universe, and it initiated further new star births. These early supernovas produced the carbon, oxygen, and nitrogen that after hydrogen and helium are the most frequent elements in the Universe. Four of these five (H, C, O, N) make up 98 percent of the mass of living organisms. The first series of star births was followed by a second one, providing most heavy elements, and then in a third series our own Sun was born. Presently this third wave of star births is the main process of the Universe (Papagiannis 1984).

This most interesting last series of star births, which began 10 billion years ago, is considered slower and longer than the previous ones. Our own galaxy, the Milky Way, characterized by the presence of huge gas clouds containing primarily molecular hydrogen and 1.75 percent heavier elements, had been thrown out from the supernovas. In certain areas of the galaxy huge clouds of dust assembled. It turned out that this "star dust" is the place of very active chemical processes (Greenberg 1984). Most of the dust formed from tiny silicate grains surrounded by a mantle of various simple compounds created by photochemical reactions. These are just the kind of chemicals important for the origin of life: water, carbon monoxide, carbon dioxide, ammonia, various cyanides, simple sugars, amino acids, etc. (Irvine and Hjalmarson 1984). The amount of the organic material thus formed is huge, an estimated 0.1 percent of the Milky Way's total mass.

There are exciting theories that assume a very important role for "star dust" not only in the origin of life but in the birth of present-day stars (Gehrz et al. 1984). The densest clouds are the most active regions of ongoing star formation. For reasons not yet fully understood, the stellar clouds occasionally undergo a gravitational collapse in which huge masses of dust form high-density protostars. The condensed material gets hot very quickly, and in certain stages nuclear fusion reactions start with hydrogen burning, which is the sign of a new star being born. In most cases the birth of a new star is accompanied by the formation of planets that also originate from condensed star dust. First, the protostar is surrounded with dust particles, and when the nuclear burning of the star begins, the dust is heavily irradiated by the star's radiation. Some of the organic substances formed evaporate and are destroyed; some others undergo further photochemical reactions, probably resulting in rather complex organic compounds. In the distant dust local condensation may begin, and meteorites and comets of various sizes (from centimeters to several kilometers in diameter) are formed. Planets are created by the further condensation of meteorites. It is most important that the new planets are subjected to the shower of organic dust and meteorites for millions of years following their birth (Delsemme 1984).

Enormous amounts of water and organic materials flow to the surface of the new planets. In other words, all basic compounds necessary for the origin of life are available in great quantities and high concentrations. Of course, life does not start on all planets. Planets too close to the central hot star lose their water and atmosphere very soon because of high temperatures, while others get too far from the central star and lack the radiant energy needed for starting life. Very probably Earth is in the "biogen zone," as is shown by the evolved biosphere (Oro et al. 1982). We can deduce from the above facts that the origin of life is not an unusually rare phenomenon but an essential part of the Universe's evolution.

Several scientists have emphasized that the eventual structure of our Universe is very improbable. Even small changes in the initial conditions could have produced major effects (absence of hydrogen, very short-lived stars, absence of heavy elements in interstellar space, etc.), which would have precluded the evolution of life. These considerations have led to the formulation of the "Anthropic principle" (Papagiannis 1984), which states that "our presence specifies the basic properties of the Universe." This is a beautiful and romantic idea; it would be unwise to pry into its firm scientific basis.

From the heavens, let us turn to the Earth again. Building blocks of the biopolymers are formed not only in interstellar space; conditions for their synthesis also were provided on primeval Earth. Photochemical reactions of simple building blocks H_2O, CO_2, N_2, H_2CO, and the like were examined in great detail (Canuto et al. 1983), and conditions for the synthesis of amino acids, urea, and cyanide derivates were found (Schlesinger and Miller 1983a, 1983b). Prebiotic chemistry of nucleotides is also well established (Ferris et al. 1984). In the light of these results the assumption that the atmosphere of primeval Earth was the place of very active chemical processes seems to be proved. The simple organic compounds that are necessary for the synthesis of the biopolymers (proteins, nucleic acids, carbohydrates) of living organisms very probably formed by ultraviolet radiation, which was much more intense at that stage of the Earth's history. It is more difficult to reconstruct the abiogenic synthesis of biopolymers themselves. Until now students of the "origin of life scenarios" were compelled to ignore some important conditions, and their primary aim seemed more logical beauty than criteria of proof (Kuhn 1976, Fox 1980, Olson 1981, Matsuno 1984).

What is the problem? Even a most simple self-sustaining organism, a bacterial cell, continuously exchanges energy and material with its environment. Its metabolism is controlled by enzymes—biocatalyzers of living organisms. The cell produces specific biopolymers, and one of these, the DNA, controls the structure of proteins through the genetic code and, with the contributions of specific enzymes, is able to self-replicate. The whole cell is an autonomous unity, an individual *system* that is able to *replicate in time and space*. All of these properties are based on a few thousand highly organized chemical reactions.

The problem, therefore, is to reconstruct the developments that have led from the abiogenic synthesis of the building blocks to the replicating cell.

The thermodynamic conditions of molecular evolution

In the last thirty years thermodynamic models have been worked out for studying the initial conditions of evolution (Prigogine 1955, Katchalsky and Curran 1965). From the energetic point of view, a system can be characterized by the following equation

$$F = E - TS$$

where F is free energy, E is total energy, T is absolute temperature, and S is entropy (the measure of disorder). In an isolated system entropy steadily increases until it reaches maximum. Accordingly, an organized structure cannot evolve spontaneously in isolated systems. Living organisms are characteristically open systems and perform their vital functions in open systems. Open systems are characterized by a constant exchange of matter and energy with their environment. When thermodynamic equations are applied, it becomes clear that the open system has a constant entropy exchange with its environment. The entropy exchange of an open system within "dt" time is

$$dS = d_eS + d_iS \qquad d_iS \geqq 0$$

where d_eS is the entropy flow between the system and its environment, and d_iS is the entropy production of irreversible processes within the system (diffusion, chemical reactions, heat conduction, etc.). In isolated systems $d_eS = 0$ and $dS = d_iS \geqq 0$, d_eS can be equal to, smaller, or greater than zero as a consequence of the constant exchange of matter and energy with the environment. Therefore, there are situations in which a system's entropy is smaller than it was in its original state. This situation can be sustained for a considerable time if the following condition is satisfied:

$$d_eS = -d_iS \leqq 0$$

That is, the situation can be sustained if its entropy production is continuously transferred to the environment. Such systems are called *steady-state systems*.

The simplest model of steady-state systems consists of an energy source, an energy "sink," and a "system" connecting the two, the thermodynamic characteristics of which are to be studied. The energy constantly flows through the system from the source into the sink, and after a while a steady state develops. The parameters of the energy flow through the system become constant. As a general law, such systems are not in a state of thermodynamic equilibrium and therefore are not characterized by an entropy maximum. On the contrary, the energy passing through the system "organizes" it, thus reducing its entropy.

Let us consider the case of a container filled with ideal gas and divided into two parts by an adiabatic, porous wall. One of the parts is connected with a "sink" at a lower temperature, the other with a "source" at a higher temperature. It can be demonstrated that as long as there is a difference in temperature between source and sink, and a constant energy flow through the system, a difference in concentration appears between the two compartments of the system:

$$\text{if } T_1 > T_2$$
$$\text{then } c_1 < c_2$$

(For a detailed discussion, see Morowitz 1968.) The entropy reduction caused by the temperature difference lasts as long as the steady state does, that is, while the energy flow continues. The system's internal organization may grow in proportion to its degree of deviation from thermodynamic equilibrium. In states far removed from equilibrium, the system may reach an inner stability that is characterized by a minimum entropy production and a maximum of free energy dissipation (Prigogine and Wiame 1946). The steady state characterizing this stability is capable, to some extent, of compensating for the system's inner fluctuations. However, should the fluctuations exceed a threshold, the system becomes reorganized and reaches a new stable state. Prigogine termed the inner structures corresponding to the respective stable zones "dissipative" because their inner organization is capable of maintaining within the parameters of the stable zone a minimum entropy production—that is, to pass the excess of entropy to the environment. The structures of living organisms are all dissipative (Prigogine et al. 1972).

Behavior of simple chemical systems relatively near equilibrium can be described by linear differential equations. The behavior of systems far from equilibrium becomes more and more complex and is characterized by nonlinear interactions and, in certain situations, by chaotic behavior. In many cases if a system deviates even further from equilibrium by taking up energy, chaotic behavior ceases and some kind of orderly behavior is manifested (Prigogine and Stengers 1984). Prigogine's famous slogan "order from chaos" was born in connection with these phenomena and unfortunately has led to much misunderstanding in biology. The equations of irreversible thermodynamics present exactly those *general conditions* that are necessary for any living system to exist. These are valid and important laws for biology like the law of energy conservation. Nevertheless, neither the law of energy conservation nor irreversible thermodynamics explains the structure of biological systems, the special causes of their origin, functioning, and complex organizations. These laws express the necessary but *not sufficient* conditions for the origin of the living system. By accepting the Prigogine school's arguments, one could not distinguish the complexity of a living cell from the orderliness of a simple chemical reaction, e.g., a Belousov-Zhabotinsky reaction. They state that in both cases dissipative structures are at work. The flaw in this argument is its confusion of *organization* and *order*, a difference we already have discussed (see also Smith and Morowitz 1982). Order in a physical or chemical system appears because of the action of microforces working between atoms or molecules, just as the orderliness of a crystal is formed in a saturated solution. A crystal is an orderly structure; its entropy is lower than that of the solution at the same temperature, but it is definitely not an organized structure. Organization never appears by leaps and bounds, and it is not the immediate consequence of the action of molecular forces. Forces acting in biological organizations are expressed in nonrandom specialized structures, and these structures are *specific constraints* that cannot be deduced from the equations of thermodynamics.

In a living cell several thousand chemical reactions are going on simultaneously, each governed by the laws of reaction kinetics and catalysis. But if we could calculate the fate of the system based on the equations of chemistry we would find the system on the way toward destruction. In spite of those imaginary calculations, cells are stable systems existing for billions of years without interruption.

The cell is a stable system because it is *organized*—because its chemical reactions are *controlled*. When a given reaction is too fast or runs in an unfavorable direction, some kind of molecular controlling mechanism immediately intervenes, which often is not directly connected with the process controlled. A clear example is the so-called allosteric regulation of certain biochemical pathways. The final product of a complex chain of reactions inhibits the activity of the enzyme catalyzing the chain's first reaction, and so the whole series is suspended. The specific structure of this controlling process is the allosteric enzyme that represents a constraint, which follows from a higher level of organization and cannot be explained solely on the basis of thermodynamics. A living cell is an "overcontrolled" mechanism in which almost every chemical reaction is controlled. Thermodynamics cannot deal with the phenomenon of cellular regulation, despite the fact that boundary conditions represented in the equations of thermodynamics are valid for controlled molecular systems as well. In many cases, though, a biological organization harnesses a given chemical or physical phenomenon, and, taking this reaction without considering the organization, we may find simple thermodynamic laws at work. Prigogine and his coworkers analyzed the process of cell aggregation of slime molds (*Dictyostelium discoideum*) and found a specific oscillation in the migration of individual cells, which they considered analogous to the Belusov-Zhabotinsky reaction. Although oscillation goes on in both cases, aggregation of slime molds is not a consequence of the system's thermodynamic parameters but of special organization, as was argued correctly by Ferracin (1984). If we take a reaction from the context of metabolism, we can certainly show that the law of chemical equilibrium is valid, but it will never explain the organization of a cell. The law of chemical equilibrium is only harnessed by the special constraints of the cell (to quote Polányi again).

It is interesting that Prigogine himself emphasizes the specific nature of systems far from equilibrium and states that there are no known universal laws describing the behavior of such systems (Prigogine and Stengers 1984, p. 144). However, he lets himself be wooed into supporting the view that irreversible thermodynamics is a final principle, valid not only for the organized cell but for ecosystems and human societies.

In my opinion, processes of living cells and organizational levels above it can be described and understood only with the help of an appropriate information theory able to deal properly with the problem of organization.

Again, I would like to emphasize that I do not deny the use of irreversible thermodynamics within its proper limits, or belittle the merits of the Prigogine school, I merely oppose an irrelevant extension of the validity of thermodynamics.

Returning to simpler chemical systems, we must briefly discuss the characteristic processes of simple systems that are far from equilibrium and lacking specific organization. There is a very important feature of steady-state chemical systems far from equilibrium: it can be shown that in these systems a cyclic flow of matter may begin (Onsager 1931). Taking a chemical system of three species of molecules, there are three possible chemical reactions:

$$A \Leftrightarrow B, \quad B \Leftrightarrow C, \quad C \Leftrightarrow A$$

At thermodynamic equilibrium there is no material flow because all reactions are reversible and rates of chemical change are equal in both directions. In an open system, however, the situation is different. Suppose that one element of the system is capable of the following reaction:

$$A + hu \rightarrow B$$

In the case of an adequate, continuous energy influx the system absorbs energy, moves away from the state of equilibrium, and a net cyclic flow of material with the direction of A → B → C → A lasts as long as energy flows through the system.

An essential aspect is that in the course of moving away from equilibrium, the system absorbs energy, which implies the *storage* of energy. The mode of storage depends on the system's characteristics, i.e., on the quality of the source and sink, and on the quality of the structure connecting the two. For example, a monoatomic gas can store energy only by producing charged electrons and ions, and only for a very short time. However, a complex chemical system is capable of storing energy in numerous ways (in covalent bonds, ionic bonds, weak interactions, etc.). The *stability* of these bonds is crucially important; the longer the life span of bonds, the more energy can be stored. By producing more complex molecules, the system itself becomes more complex, and its *orderliness* increases at the expense of energy flowing through it. The level of order is proportional to the free energy content of the system. The order of a system also is influenced by the nature of the energy source. It has been shown that in a system containing the basic bioelements, i.e., C, N, H, O, P, and S, it is the energy of visible and ultraviolet light that can produce a maximal level of order. According to Morowitz's (1968) calculations, *the degree of order is proportional to the mean residence time of the energy in the system* while flowing from the source toward the sink. The more organized the system, the longer it stores the energy. Convincing proofs of this idea also are provided by modern cell biochemistry (Welch 1977).

Today, just as at the time of the origin of life, the Earth's surface is a complex, energetically open system, showing a constant energy exchange with its environment. It absorbs the radiant energy of the Sun, and this absorbed energy is radiated into space at other frequencies. On the basis of the above-discussed thermodynamic model, the energy flow going through the Earth theoretically allows the system formed on Earth to move away from thermodynamic equilibrium, to increase its inner complexity and order through a succession of stable steady states, and to develop dissipative structures, with the result that various kinds of *organizations* emerge.

Here again we raise the question of complexity and order, because the lack of distinction between these two concepts is the main cause of theoretical problems in studying the origin of life. The appearance of both complexity and orderliness is characteristic of the structures of living systems. The average distribution of amino acids in the proteins of modern organisms is very close to *random*. This simply means that, in general, proteins are complex structures because in their description we need exact sequences of amino acids and we cannot apply rules to simplify this description. We also know proteins with highly ordered structures. In the keratin of human hair various amino acids are connected in a given, repeated order. The existence of numerous *copies* of a certain complex molecule in the cell also can be seen as a kind of order because the state of disorder would be characterized by a random distribution of the various sequences, and the chance that a given sequence is represented by two or more copies is rather small. The possible number of sequences is on the order of 20^{500}. (The number of atoms in the Universe in hydrogen equivalents is approximately 10^{80}, which gives an idea of how large the former number is.) The fact that amino acids of the cell system form giant molecules also represents a kind of order because entropy and correspondingly complexity would be even higher should all amino acids be monomers or their atoms parts of simpler molecules such as ammonia, water, etc. Therefore, the state of an open system can be characterized by the simultaneous effects of two well-distinguished tendencies, i.e., the growth of complexity and the emergence of order in certain parts or parameters of the system.

In a closed system processes move in the direction of growing complexity (disorder), while in open systems the energy flow would enhance order in certain parts of the system at the environment's expense. The most important thing is that in a simple, open chemical system the emerging complexity is always *unorganized;* i.e., the emergence of complexity per se can be explained by thermodynamic laws, but in the case of an organized system we need additional explanations for organization. Organization can be understood only if we have an explanation at two connected levels. Thermodynamics provides explanations only for the lower level; it explains how complexity arises from the interactions of the molecules; while at the higher level we need an explanation for the existing constraints that form the given organization. This explanation is not provided by thermodynamics. In most cases these constraints include a certain order, for example, a great number of copies of the same polymer or a given structural arrangement, etc. Therefore, organization is characterized by the *mode* of connection of complexity and order, and thus organization cannot be described on the basis of irreversible thermodynamics alone.

Summarizing the characteristics of a steady-state chemical system far from equilibrium, we can state:

(1) Entropy production is lower and the level of free energy is higher than in an equilibrium system at the same temperature.

(2) With the advance of time, the molecular complexity and free energy content increase.

(3) With the advance of time, order increases. In an unorganized system the degree of order is roughly proportional to the energy-storing capacity and the mean residence time of energy in the system.

(4) The relationship between the order and complexity emerging in the system depends on the system's parameters and follows unequivocally from its microstates.

(5) Cyclic material and energy flows begin in the system.

(6) A continuous energy flow is necessary to the steady state's maintenance.

The chemical composition of the evolutionary system on Earth

Since there are quite good estimates available for the possible composition of the Earth's primitive atmosphere, the nature of processes that took place within it can be modeled, and the results can be tested in laboratory experiments. Generally, it is assumed that the system contained carbon, hydrogen, nitrogen, and oxygen in a certain proportion. Knowing the respective chemical parameters, the equilibrium concentration of each simple molecule can be calculated (Dayhoff et al. 1964). According to these calculations, concentrations of CO_2, CH_4, H_2O, and N_2 were highest, and those of molecules of biological relevance were extremely low because producing these molecules requires a great deal of energy. For example, in a system containing C, H, N, and O in a 2:10:1:8 molar ratio, the equilibrium concentration of acetic acid is 10^{-9} mol/l and that of glycine is 10^{-21} mol/l. This means that in a thermodynamic equilibrium at terrestrial temperature, without an external source of energy, the system could have produced only the simplest molecules.

Computer simulation has shown that in a steady-state system increases of the free energy content were followed by an increase in the *variability of chemical composition* (Morowitz 1971); that is, the complexity-enhancing effect of energy flow is well supported. It also was shown that this increased variability would still include compounds important for life.

However, the atmosphere of the primitive Earth, having constantly been exposed to the irradiation of the Sun, was not in a state of equilibrium. According to the model experiments of Groth (1960) as well as those of Noyes and Leighton (1941), in a mixture containing CO_2 and H_2O the ultraviolet irradiation initiates the following overall reaction:

$$CO_2 + H_2O \gtreqless H_2CO + O_2$$

That is, carbon dioxide and water may react to produce formaldehyde and oxygen, and the reaction of the latter two produces carbon dioxide and water again. This reaction is the basis for biological photosynthesis and for the ecological carbon cycle. During photosynthesis the carbon dioxide turns into more complex aldehydes and carbohydrates that enter the metabolism and finally are oxidized into carbon dioxide and water. Formaldehyde produced in the carbon dioxide–water system is a compound of high reactivity. It can react not only with oxygen but can take part in the synthesis of other carbon compounds. As a consequence of irradiation, the carbon dioxide–water mixture becomes a system of a higher free-energy level, and its chemical potential also increases. Naturally, after a period of time the material cycle closes, and the system emits the absorbed energy in the form of heat, while the initial simplest compounds are regenerated.

Under the influence of these model experiments, solutions more closely resembling the elemental composition found in living organisms have been studied extensively (containing phosphorus and sulfur in addition to carbon, oxygen, and hydrogen). As shown by several experiments, when such a CHNOPS solution is irradiated with ultraviolet light, a series of photochemical reactions occur, and numerous compounds of medium molecular weight are produced. Some of these compounds have a short life span, and they dissociate or change, while others with a longer life span become concentrated. Under the influence of continuous irradiation, the system shifts toward increasingly higher energy states; compounds of higher and higher free energy content are continuously formed in it; and the very molecules (amino acids, organic bases, sugars, etc.) that are the building blocks of the molecular structures of living organisms appear (Fox and Dose 1972).

Let us examine this system from the point of view of the material cycles involved. For the sake of simplicity let us consider only the CHNO elements. In a state of equilibrium the system contains only small molecules such as CH_4, H_2, NH_3, etc., in considerable concentration. The set of compounds containing three or four atoms can be regarded as a material source and demonstrated as a cube:

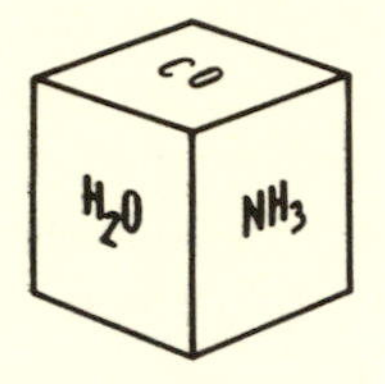

Figure 6

As a consequence of irradiation, photochemical reactions ensue, and more complex compounds are produced. Let us pick out a single chain of reactions and mark each reaction product by a letter of the alphabet:

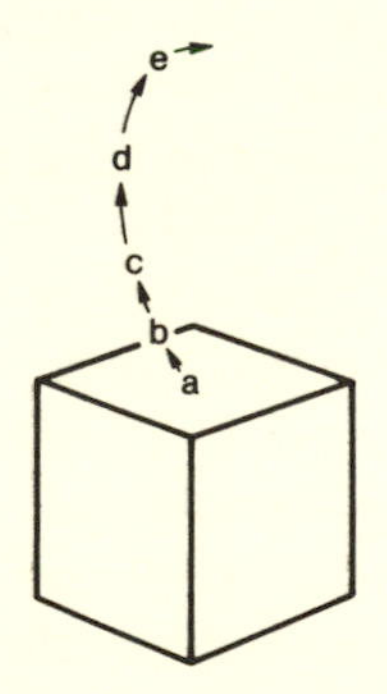

Figure 7

As there is no particularly stable compound among those produced in the CHNO system, we can safely state that these compounds would—in one way or another—be transformed into the original basic molecules of the material source:

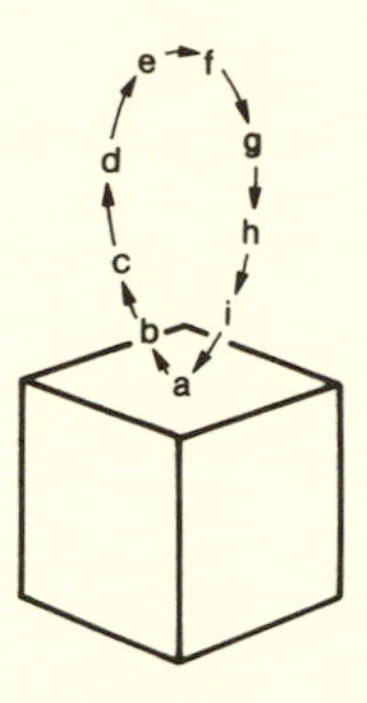

Figure 8

This means that a closed network of processes develops, through which the atoms are driven by the free energy absorbed by the system. Naturally, not only a single cycle but a complex diverging and intertwining net of cycles develops. Marking the individual compounds by dots, it can be schematically visualized as follows:

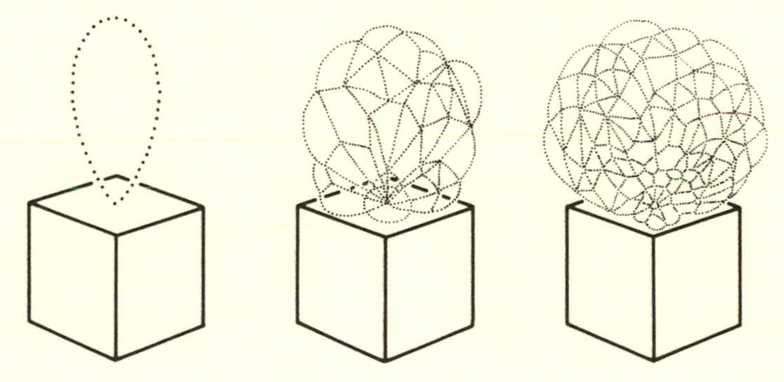

Figure 9

In this system no reaction can occur that would not take part in the continuously flowing material cycle. Undoubtedly, only a small fraction of the theoretically possible reactions are realized. The thermodynamic models determine the system's limiting energetic conditions.

The most important of these conditions is that the system may get into a steady state in which it has minimum entropy and maximum free-energy content. From the point of view of evolution, the crucial question is what kind of organization will characterize this network of processes.

Reaction parameters characterizing the evolutionary system

Let us consider the CHNO system again and divide it into two subsystems, one consisting of small molecules denoted by "S," the other of more complex molecules, generated from the small ones, denoted by "M." The connection between these two subsystems is demonstrated by the scheme

$$S + \text{energy} \Leftrightarrow M$$

By absorbing external energy, a continuous material flow is established between S and M, and the free-energy content of M increases. The free-energy increase of M depends on the material flow's mass and the specific energy-storing capacity of the molecules. Let us examine the factors determining the free-energy increase of M:

The mass of material flow is determined by the parameters of the energy flow and the equilibrium constants of various chemical reactions. Generally, these constants are such that the relative concentrations of the simpler compounds are much higher than those of the more complex ones. This is well demonstrated by Morowitz's data on the distribution of carbon on the surface of Earth (table 1):

Table 1 Distribution of terrestrial carbon (after Morowitz 1968)

Compartment	Total carbon $\times 10^{18}$ g
Carbonate in sediments	67,000
Organic carbon in sediments	25,000
CO_2 in the atmosphere	2.35
Living matter on land	0.3
Dead organic matter on land	2.6
CO_2 of dissolved hydrocarbonate in oceans	129.7
Living matter in oceans	0.03
Dead organic matter in oceans	10.0
Total	92,144.98

Huge material cycles characterize the developed biosphere, and the chemical elements of which living organisms are composed also are parts of the material cycles. It is increasingly clear that the most important characteristic of life is the complex network of joint material cycles (Mizutani and Wada 1982).

The total biomass (the living and dead organic matter of oceans and lands) is 12.93×10^{18} g; the carbon in this is hardly more than 0.01 percent of the total carbon store. In the present stage of evolution, as far as material cycles are concerned, a steady state has certainly been attained. The carbon flow per year also has been determined (Hutchinson 1954, cited by Morowitz 1968), and the following values have been obtained:

Amount of CO_2 fixed on land: $0.073 + 0.018 \times 10^{18}$ g/year
Amount of CO_2 fixed in oceans: $0.43 + 0.3 \times 10^{18}$ g/year

These calculations of the actual material flow assume equilibrium, but the central characteristic of the terrestrial system is its nonequilibrium nature. This also implies that the rates of reactions participating in the material flow play a limiting role in determining the actual flow. Among synthesis reactions, autocatalytic reactions have the greatest reaction rates. It is easy to demonstrate that if other conditions for their existence also are given, in a steady-state chemical system these reactions will predominate (King 1977, 1982).

The free energy enters the system through photochemical reactions; thus the nature of these reactions is of primary importance in creating the bulk of material flow. The rate of reactions subsequent to the photochemical reaction also is significant. The reaction rates are the decisive factor in determining *the time that energy spends in the system*. Those reactions that "remove" the material relatively rapidly from the system and recirculate it to the source of basic compounds decrease this value. Those reactions, however, that drive the material to byways keep the energy in the system longer and consequently increase the total free energy of subsystem M, provided that the mass flow rate into M is constant. A kind of "selection" automatically starts among the various types of reactions in the system. All changes tending to increase free energy's storage are incorporated into the system and become a stable part of it. From the energy-preserving reactions the ones having greater capacity will be *selected*. These are the autocatalytic reactions since their rate can exponentially increase in time. On the basis of King's (1977) investigation it also can be stated that in a sufficiently large steady-state system, with conditions adequate for various reactions, the different autocatalytic reaction chains may interlock and produce increasingly greater *autocatalytic complexes*. If two autocatalytic complexes "compete" for initial materials, the one with a more compact physical structure, and consequently with more catalytic sites per unit of surface area, has a selective advantage. Such catalytic particles ensure greater reaction rates. This so-called "equilibrium" effect was first described by Martell (1968, cited by King 1977). Autocatalytic systems have "self-regulating" features that make their relatively longer life span possible. This coincides with Ashby's criteria concerning the stability of different regulatory systems: those systems are stable in which the "groups" are closed toward transformation, i.e., in which states are cyclically repeated (Ashby 1956).

Stability problems in complex chemical networks were investigated by Matsuno (1978), who found that the probability of forming cyclic processes was rather high, and these were especially stable kinetically. He also proposed that *compartments* spontaneously emerge within the reaction networks, which have a tendency to develop self-preserving organizations (Matsuno 1980, 1981). Matsuno made a detailed study of the kinetic behavior of compartmentalized polymers. He found that with time the ratio of *stable polymers* increases because of simple kinetic relationships. He also found that the mean decay time of the polymers within the compartment continuously and irreversibly decreases (Matsuno 1974, 1977). Smith and Morowitz (1982) state that reaction networks emerging in the CHNO system are far from random; they show signs of definite organization.

In summary, the M subsystem's free-energy increase is decisively influenced—with regard to material flow—by energy-storing and preserving cyclic, autocatalytic reactions.

As to *the specific energy-storing capacity of the molecules in the M subsystem*, in the CHNO system the compounds of low molecular weight can store relatively little energy because the H–O, H–N, H–C, etc., bonds at the chain ends of the molecules have small energy content. As the proportion of C–C, C=C, C–O–, C–N–, etc., chain-elongating bonds grows, the average molecular weight in the system increases, and the storage of an increasingly greater amount of free energy becomes possible (Morowitz 1968). Consequently, the CHNO system shifts toward synthesizing giant molecules to reach the free-energy maximum.

The two important free-energy increasing tendencies, therefore, are autocatalysis and the size of molecules. The latter, if only the covalent bonds are considered, has a definite optimal upper limit. For molecular weight over a few million, the molecules are extremely sensitive to simple physical effects; they readily disintegrate. However, if not only covalent bonds but the variety of *functional connections* involved in the energy interaction are considered, not the giant molecules but rather structures above molecular level—such as cells, organisms, ecosystems, etc.—represent the upper limit for the size of structures produced in the system since cohesion and bonding energy also manifest themselves in these higher structures. With this addition a new assumption can be proposed: the M subsystem reaches its maximum free-energy content when all of its atoms become part of a system produced by an autocatalytic process and acting as a single functional entity. This structure will be a dissipative one and, if the parameters of the external energy flow remain constant, can be stable in the state of minimum entropy.

Information parameters characterizing the evolutionary system

Different energetic considerations, especially the models developed by the Prigogine school, specify the energetic boundary conditions of the evolutionary system on Earth as well as of systems in general that are capable of evolution. If an evolutionary system develops at all, it can function only within these boundary conditions. Unfortunately, these models do not answer whether an evolutionary system's emergence is indeed a necessity or only a phenomenon occurring under very specific circumstances. The question of the direction of evolutionary changes, apart from the changes of energetic states, also arises. Still more important, the nature of the *organization* of the complex system appearing in the course of evolution is still an open question. We have made a few steps toward understanding these questions—though not ultimately resolving them—while examining the reaction parameters. In those complex systems developing in the course of evolution, a definite tendency for autocatalytic processes to prevail and for systems to become organized as functional units seems evident. A still better understanding of the problem can be expected by studying the information parameters of increasing complexity in the evolutionary system.

The evolutionary zero-system

One of the living organism's specific features is the continuous (assimilative and dissimilative) chemical activity called metabolism. It is obvious from previous sections that the development of complex reaction networks preceded the emergence of life. Evolution started when conditions for development of a steady-state system had been established on Earth. The proliferation of compounds, starting with irradiation of the CHNOPS system, was the first step of deviation from equilibrium and of *evolution*.

The concept of the zero-system of our model was introduced earlier. This is an intuitive concept, reflecting the state of a system exclusively characterized by thermodynamical parameters. A natural zero-system does not exist, of course; at most, there is only a loose approximation of it. Nevertheless, by using the abstract concept of a zero-system we try to trace the emergence of organization and function.

From an organizational point of view we call the zero-system the "ground state" of the molecular system immediately preceding the start of evolution on Earth. The molecular zero-system should (a) contain a sufficient variety of reactive atoms that form chemical bonds (CHNOPS supplemented with metals, halogens, etc.), (b) have a temperature that never exceeds a limit (certainly below 100 C), and (c) be charged continuously with energy by the Sun's radiation. Conditions on primeval Earth seem to have satisfied these criteria. It can be asserted on the basis of the laws of chemical kinetics and thermodynamics that in such a zero-system a series of chemical reactions immediately begins: more complex molecules are synthesized, material cycles develop, autocatalytic reaction chains appear, and molecules of a higher level of complexity (primarily various polymers) are formed.

Although it is inferred from the zero-system's thermodynamic parameters that synthesis of more complex molecules should start and that the system should contain increasingly complex structures, this does not allow us to draw definite conclusions about the specific features of molecular structures.

If in a system that synthesizes macromolecules the produced molecules do not influence the probability of genesis of subsequent molecules, the structure of the molecules formed (the sequence of building blocks) will be determined exclusively by the system's parameters, i.e., temperature, energy relations, catalysis, etc. Therefore, the information content of these structures is called *parametric information*. It is known from model experiments that artificially synthesized proteins, with random distributions of amino acid sequences, show a divergent variety of catalytic activities (Roheting and Fox 1967). This activity is based on the chemical features of the amino acids, which themselves are capable of increasing the rate of various reactions (Capon 1964). This catalytic activity feeds back to the producing system and considerably influences the structure of the giant molecules produced. For example, it has been shown that from amino acids condensing at higher temperatures, proteins of certain specific sequences are produced instead of ones with random distribution (Fox and Dose 1972). This effect, which manifests itself in structure-shaping, will be called the *protofunction* of a given compound.

According to our definition, a zero-system has no organization; therefore, there are no functions or components in it in the sense of our definition. But with the onset of molecular interactions, we have to consider these transient interactions. Our definition of function, however, is inseparable from the system's organization, as molecular interactions are regarded as functions only if we already have a second level of event-space. Until we can define the system's organization, we cannot define functions. This chicken-egg problem can be solved by introducing the concept of protofunction. Protofunctions are concerned with neither the organization of the system nor with the system as a unity, but with the interactions of individual components.

If in a system that synthesizes macromolecules the physical parameters result in polymers with randomly distributed monomers, then there are obviously no protofunctions of any kind. But if one of the polymers influences the incorporation of monomers, and as a result new kinds of polymers are produced with structures that can be distinguished from the previous generation, then the action of a protofunction is manifested. The effect of a compound on the probability of genesis of another compound, without any particular structural changes, also is considered a protofunction.

In a system synthesizing macromolecules of the slightest degree of complexity, protofunctions necessarily appear without requiring the fulfillment of any specific conditions. For example, reactions leading to the formation of complex molecules quite often feed from a common pool of building blocks, and it is exactly this common pool that influences the structures and probability of the genesis of formation of complex molecules. That is, protofunctions emerge spontaneously. A similar finding has been published by Pattee (1965), who calls this period of evolution the phase of "molecular automata."

If the original parameters promote the formation of structures constructed from building blocks in a random distribution, the protofunctions are formed at random even in the zero-system. Each protofunction starts to act as certain compounds appear, and they later disappear without trace; i.e., initially the protofunction content of the zero-system *fluctuates at random*.

We have arrived at the most important part of our argument: in a zero-system the functional organization of the formation and decay of components immediately starts when an *organizing agent* appears. For example, if we place a single suitable bacterium into a "system" consisting of water, some inorganic salts, and simple organic compounds, in a couple of hours the system will be transformed; it will contain a huge mass of bacteria with complex, highly organized organic materials. These changes occur at the expense of energy sources of our system, and all laws of thermodynamics prove to be valid. Nevertheless, it is a fascinating change brought about by an organizing agent of some cubic microns in size.

From the organizational point of view this change is very simple: the organization of the organizing agent itself is multiplied by billions. That is, replication in space occurred; the organizing agent simply replicated itself. An individual bacterial cell can spread its own organization by producing its own components in a large number of copies at the expense of the system's material and energy sources.

The basis of its organizational effects or, to be precise, the necessary and sufficient basis is *replication*.

The bacterial cell is a perfect organizing agent, but we cannot explain the start of evolution by the sudden appearance of a bacterial cell because our aim is to explain the very appearance of such a highly organized *system*. This example, however, may help us work out a useful analogy. The concept of the *system precursor* has been mentioned. It is a minimal network of components that is able to maintain its own organization and also to transform an unorganized system into one of similar organization. By this criterium a bacterial cell is a system precursor, although it represents not a minimal but a complex organized network. Our task is to find or define the simplest possible system precursor that is able to organize a zero-system through the appearance of protofunctions and to reasonably establish its *spontaneous appearance* in the zero-system of Earth.

Reproductive chemical networks

In the previous sections we presented in detail the notion that in an open chemical system material cycles "energized" by the energy flows through the system would begin. We also have shown that in these systems only *cyclic* chemical reactions can exist for an extended time because any others quickly use up their supply of reactant compounds. Conditions for the emergence and existence of completely recycling chemical networks have been analyzed by King (1982). He found that if a reaction network includes at least two bimolecular reactions, complete *chemical reproduction* may emerge. It also is evident that the *reproductive chemical network* (RCN) has to include reactions connected with processes that provide energy to maintain the material flow. Such a cyclic chemical network is not very complex, and King (1982) provides reasonable proof of its *spontaneous appearance*.

A network suitable for chemical reproduction still may be open. Every reaction that takes material away from the network will destroy it, but reactions that join the chemical reproduction become part of the network, which is enlarged, though its organization remains unchanged (figure 10).

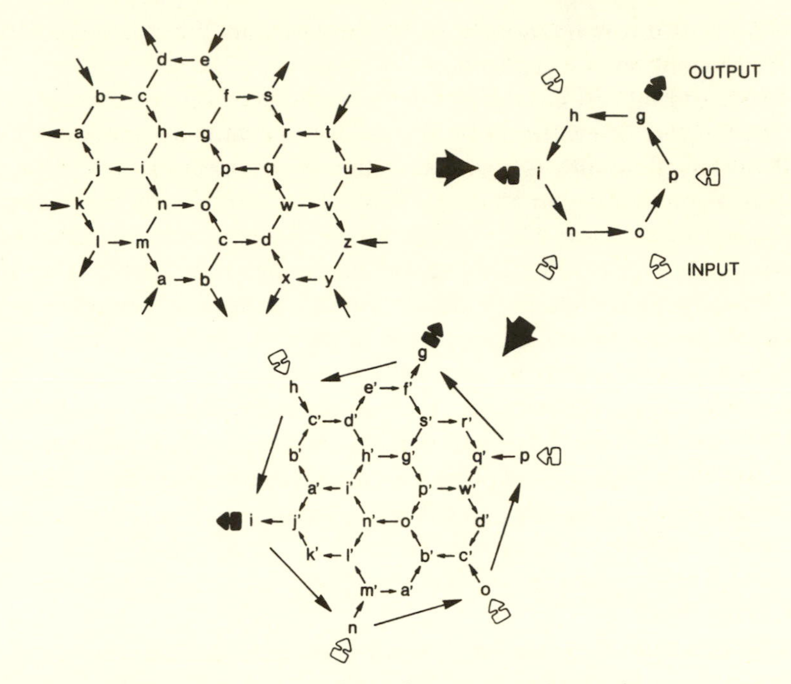

Figure 10 Spontaneous emergence of reproductive chemical network.

With the spontaneous appearance of reproductive chemical networks, *selective processes,* which are a new feature of the system, also emerge. This selection is not Darwinian or competitive, which is well known in biology. The reproductive chemical network as a *system,* as a proto-organization, will select those processes that will not destroy the system. The environment of an RCN "offers" various substances and possible chemical reactions, depending on its parameters and composition. Reactions that fit into the RCN without destroying its cyclic reproductive organization can be incorporated, while others, which can effectively obstruct reproduction, will destroy it.

The appearance of this kind of selective mechanism will make the RCN very sensitive to its environment, especially if there is only a single reproductive chemical network. Of course, it is possible, but quite unlikely, that the zero-system on the surface of Earth was a stable single unity only because of its sheer size. It is much more likely that the chemical zero-system *compartmentalized* before reproductive organization appeared.

Such a compartmentalization process was analyzed by Matsuno (1980) on theoretical grounds. It also was shown empirically by various experiments that in the primitive ocean of Earth, "microspheres" or "protocells" may have been formed, for example, by random polymerization of simple amino acids (Fox 1973, Yanagawa and Egami 1980). These microspheres could have been suitable carriers of appropriate RCNS. Koch (1984) elegantly proved that without compartmentalization, a replicative system is unable to evolve. It also is quite evident that in the zero-system the process of compartmentalization created an enormous amount of *unorganized complexity* that was necessary for further evolution.

In his work on self-reproducing automata von Neumann (1966) suggested that there was a critical level of complexity below which a system would fall apart because of the environment's disturbing effects. Alternatively, above this critical level the system's complexity may increase. Only one microsized compartment is very likely under the critical level of complexity, but the whole zero-system of Earth is certainly above it.

The idea that the emergence of RCNS occurred in microsized compartments is well supported. If RCNS were localized in tiny compartments, then they also could have existed simultaneously in great numbers ($10^{30} - 10^{40}$). Thus besides the above-mentioned organizational selection, competitive Darwinian selection processes started to operate, which could strongly promote the system's evolution. For the simultaneous emergence of both these mechanisms, a system of sufficient complexity is required.

One interesting question: is the primitive proto-organization expressed in the RCN the only possible type of organization, or are there a variety of other organizational types that are theoretically or empirically possible? The answer seems simple. The organization of RCNS is a reproductive organization by definition. It is the organization of *being;* therefore, it precedes any other kinds of organization. Any organization that cannot maintain itself is only temporary and irrelevant to evolution. The meaning of the concept of organization is relevant only if there is a *teleonomic* goal or function to which organization can be assigned. A complete description of organization, therefore, must describe some goal or function. As a consequence, the descriptions of apparently simple organizations are highly complex because they can be represented only on multilevels. The goal or function of RCN's organization is in *itself;* it can be expressed by the verb "to be." *Any concrete mechanism* that is capable of "being" is suitable for the given goal. Therefore, an organization of "being," except for certain limits, permits enormous variability, which is clearly showed by the rich variety of living organisms. Scientists, arguing against the idea of the spontaneous origin of life, have shown, on the basis of various statistical considerations, that the probability of an organization like that of a living cell emerging from simple compounds is negligible—practically zero (Wigner 1961). These arguments are rather uninteresting because the organization for "to be" in an RCN is not necessarily complicated; any cyclic chemical reaction is enough to create one, and if it already exists, then it can evolve very quickly helped by the promoting effect of the above-mentioned selection processes that appear in the complex zero-system. (See Kampis and Csányi 1987b for an investigation of some properties of reproductive functional networks by computer simulation.)

Replication, replicative information

The reproductive chemical network is not a component system according to our earlier definition because its constituent compounds have no functions and therefore are not components. Reproduction of an RCN is realized through the spontaneous reactions of its molecules, and the existence of the whole cycle depends on *reaction rates.* There are no controlling structures and processes in an RCN to make the "system" independent of reaction rates, which could make reproduction *replication.*

When a simple substance, a vitamin or an amino acid, is removed from a bacterial cell's environment, its growth immediately stops. But the cell's reaction network can resist breakdown and the cell can live for a long time because while "fasting" various controlling systems are activated and the whole network of the cell's metabolism is channeled into a special kind of "parking state." So a bacterial cell's system is able to modify and reorganize the behavior of its own reaction network and defend itself to a limited extent against environmental effects.

The maintenance of a network of reactions without controlling systems is based on coordination of the various reaction rates. Thus it is unable to protect itself and is highly unstable.

It was noted earlier by Pattee (1977) that a stable replicative system can emerge only if it separately contains its own description, which acts as a controlling mechanism. An RCN obviously is not such a system.

In my opinion, the major transition between chemical and biological evolution is therefore represented by changes that are manifested by the appearance of components carrying functions that also constitute a unified replicative organization, replacing the simple reproductive structure of the early RCNs and embodying the description of the system itself for control and maintenance. All conditions for such a transformation are present in the zero-system. Formation of a complex but in some parts orderly system is possible in accordance with the laws of thermodynamics. The energy flow necessary for creating such a system is available, and the spontaneous formation of RCN produces the organizational agent that is able to create an organization higher than itself through its own transformation. The reproductive chemical network therefore is an *autogenetic system precursor*.

We have to note again that the appearance of RCNs, which are just physical microcompartments of the zero-system, represents a kind of order in the system because these microcompartments are very much alike, at least at the beginning. But when they gradually undergo the strong effects of Darwinian and organizational selection processes, they become different; that is, a kind of complexity arises in the zero-system. Order and complexity arising simultaneously result in *organization*.

Organizational selection, which will be called *replicative selection* for reasons detailed below, results from the reproductive organization of the RCNs. According to King's studies, this organization may be very simple because it requires only a cycle of a few simple compounds, maintained at the expense of an outside energy source. In spite of its simplicity, however, such a reproductive cycle represents a very powerful organizing agent. It was mentioned that the various compounds of the zero-system can have protofunctions. During the creation of an RCN various compounds of the most different protofunctions may participate in the network. Compounds that through their protofunctions lead the network toward stability carry *real func-*

tions, as they have been defined in our model, and therefore these compounds become *components*, and the organization of an RCN may transform from reproduction into replication.

In addition to being promoted by replicative, organizational selection, this transformation is promoted by a competitive, Darwinian selection. In the first phase of selection the primary stability of the reaction network is the most important parameter. Each compound that enters the network and contributes to its stability through its function becomes an integrated member of the RCN. Selection of components takes place based not only on reaction rates but on a higher level of selective conditions. The condition of replication appears and acts to coordinate the functions of components, and thus it represents a primary condition for survival; i.e., it is the starting point of the new, replicative organization.

Transformation of the system from reproductive to replicative organization, transformation of the compounds selected on the basis of their reaction rates of production, and transformation of the protofunctions of the components to real functions seem to be possible, and we can assume this process is spontaneous. Most probably this transformation is gradual and is not immediately accompanied by the formation of genes and the apparatus of protein synthesis with precise replicative properties.

Behavior of the primitive RCN depends on the reaction rates of the production of its compounds. The new functional organization that emerges within it requires concrete "mechanisms" for controlling the *rates of the chemical reactions*. This was realized by the appearance of compounds with *catalytic activity*. Many proteins of modern cells are just such catalyzers. It also is evident that a catalyzer stabilizing the RCN must be formed continuously, and this can be achieved only through reactions that themselves are parts of the RCN. As soon as the RCN begins to produce catalytic components, functional information emerges in the system. The simple but highly active organization of RCN begins to transform the parametric information content of the zero-system into replicative information. With advancing time the initially 100 percent parametrical information content of the zero-system begins to decrease, and at the same rate the replicative information content begins to increase. This is the early phase of evolution's most important process.

Replicative information can be viewed as a set of prescriptions, or an algorithm, controlling the chemical reactions within the system so as to achieve the whole system's complete replication. The replicative algorithms are special constraints, formed above the level of the physical and chemical processes; they control the lower levels and harness their processes, according to the prescriptions of the new, higher level.

The ontogenetic complexity of matter is infinite and can only be approximated by models, the semiotic complexity of which can be perceived and understood by the human mind. The replicative information itself represents this kind of semiotic complexity. Therefore, the realm of biology can be completely understood.

In the course of gaining knowledge about life, knowledge of replicative information exhausts the biological information content of the system. It harnesses only a negligible part of its total ontogenetic complexity. The biological world has been increasingly separated from the physical world in the course of evolution because of the actions of replicative information. It creates its own laws in the form of algorithms subordinated to replication. Separation of the biological and physical world cannot be complete because the biological system's simplest components are physical entities. We never can understand the physical world in its totality, and therefore we never can *create* a physical world; but algorithms of biological and social systems can be completely understood. Such systems, in principle, can be reproduced in the laboratory and, if necessary, created in various new artificial forms.

For understanding molecular evolution it would be very important to reproduce the very first steps of chemical evolution in the laboratory. There are many promising experiments in this direction, and a short overview follows in the next section. We want to note, however, that experiments in the laboratory never can replace thinking and vice versa. It is a very important task of theoretical biology to clarify the problems expected to be solved by experiments; but if we have a good theory, the problem is almost solved. Without the right idea, even the best experimenter is only groping in the dark.

The phase of molecular evolution

The phase of nonidentical replication

The replicative information content of the RCNS originating in the zero-system is relatively low, practically zero at the initial moment. As soon as replicative selection and competitive selection begin to act, the ratio of replicative information starts to increase.

The replicative fidelity of early chemical systems is most probably very low. Side reactions produced by spontaneous fluctuations may "drain" the developing autocatalytic cycles, which results in the highly inaccurate replication of the whole system and its individual components. The replicative cycles are not exactly circular but rather have the character of an advancing spiral. If replicative fidelity is expressed as a numerical value between 0 and 1, in this stage it is certainly not close to 1. Thus this stage of evolution from the zero-system should be called the phase of *nonidentical replication*. Mathematical studies by Richardson (1976) showed that replication may become stabilized as an organization if its fidelity is somewhat higher than 0.5. Therefore, we may assume that the replicative fidelity of RCNS was at least this high and began to increase from this value in the phase of nonidentical replication.

Perhaps we should note here that in further discussion the concept of system occasionally will be applied to the subsystems of the zero-system—to such RCNS. This will not mean, however, that events in the subsystem can be artificially detached from events of the whole system. With the formation of RCNS two levels of events and therefore two levels of organization were created. The various thermodynamical and organizational considerations are valid on both levels. Replication of a microcompartment is understood as the replication of all components belonging to the microcompartment, but through replication of the microcompartments the whole system also is replicating. An arrangement similar to the "Russian doll" is formed. The RCN as a higher organizational unit than the participating individual compounds is a component of the zero-system, and the components of the RCNS are the components of both the RCNS and the zero-system.

A long experimental search for compounds and chemical reactions that could have been the components of reproductive chemical networks is taking place. On the basis of evolved biological systems, three lines of investigation are most preferred. Some scientists assume that even the most primitive reproductive systems must contain nucleic acids (see, for example, Eigen and Winkler-Oswatitchs 1981, Bloch et al. 1983). By simple deduction they propose that the "first component" starting evolution was a kind of TRNA, transferring amino acids to the site of protein synthesis. This approach can be questioned because, as we have discussed, it is highly improbable that a single compound could be enough to start evolution; for that to happen, an organizing agent and a chemical network are required.

In evolved biological systems specific enzymes are necessary for the replication of nucleic acids. To avoid this problem, it is frequently assumed that the primitive RNA also has had some kind of enzyme activity (Brewin 1972, White 1976). This assumption would give a simple explanation for the joint evolution of RNA's catalytic and template (self-copying) activity. Recently RNA was found in living cells that actually have some catalytic activity (Zaug et al. 1983), supporting this assumption. This line of thought also is supported by the finding of conditions under which nonenzymatic, template-directed RNA synthesis becomes possible (Inoue and Orgel 1983). On the other hand, the fact that no one has been able to produce nucleotides in aqueous solutions without specific enzymes contradicts the whole idea, and there are doubts that it can be done (Shapiro 1984).

Another tendency is to say that protein synthesis came first. This view was well articulated by Fox and his school (Fox 1973, 1980, 1984, Matsuno 1983, 1984). It is supported by the observation that proteinlike polymers can be produced by heating diluted solutions of various amino acids. The polymers form small spherules, and it is easy to imagine that these microspheres may serve as *compartments* for RCN systems. The main problem with this approach is that template-directed processes in which proteins serve as templates are not known. (See an interesting theoretical assumption by Root-Bernstein 1983.) Although simple polypeptides influence the process of their own polymerization, this is not a template mechanism. The polypeptides produced by heating contain amino acids in highly nonrandom sequences, and the individual polypeptides are built up from only a few amino acids, even when a solution of a varied composition is heated. It is quite possible that this "self-sequencing process" discovered by Fox (1984) is really an important evolutionary mechanism, but we certainly need further convincing experiments.

The third line of investigation centers on the joint reactions of nucleic acids and protein synthesis. A reaction complex is searched for which makes possible the formation of a joint feedback-controlled system for both RNA and protein synthesis (Lesk 1970, Brack and Orgel 1975, Crick et al. 1976, Olson 1981, Ishigami et al. 1984).

Characteristically the search for evolution's chemical bases misses the advantages of proper system-thinking. Most researchers are looking for concrete mechanisms, for concrete compounds, and not for conditions of reproductive and later replicative simple systems.

The way to success may be to try a deductive approach. In presently evolved biological systems there are replicative mechanisms characterized by the following conditions:

(a) In addition to replication of the whole system, specific molecules such as RNA and DNA are replicated.

(b) The replication process occurs through the action of specific enzymes, that is, protein polymers.

(c) There is feedback between the protein and nucleic acid synthesis in the form of the genetic code.

(d) For entering individual amino acid molecules into the machinery of protein synthesis, the contribution of specific RNAS is necessary.

(e) Both RNA and protein synthesis use energy that is provided by phosphorylated nucleotides that play a universal energy-transferring role in cell metabolism.

Recent results in molecular biology show that the various mechanisms mentioned above are highly complex, and it is difficult to suppose that within early reproductive chemical networks they could have appeared simultaneously in their present forms. If we look for the most simple processes and compounds connecting these mechanisms, we find there is a common compound—the adenine—participating in each of them. Adenine nucleotides are the building blocks of both DNA and RNA molecules. Various adenine derivates are coenzymes and contribute to the catalytic effects of proteins. In protein synthesis the first step of amino-acid activation is the formation of an aminoacyl-adenylate; i.e., amino acids can enter the proteins only bound to adenine. The cell's most important energetic processes are connected with adenine-containing nucleotides like ATP or ADP. Other derivates of adenine—cyclic AMP, for example—play an important regulatory role in cell metabolism. There is no better candidate for a unique compound in evolution than the adenine, and the search for the most primitive chemical reproductive systems must be concentrated on adenine derivates.

Complex chemical systems, supercycles

Even if identification of concrete compounds participating in the first reproductive chemical networks is far ahead of us, several remarkable studies deal with the organizational problems of complex model systems at the border between biology and chemistry. Among these, the most important is Gánti's "chemoton model." The self-replicating chemoton model consists of three functionally connected subsystems: an autocatalytic component-producing network, a polymer-producing network operating in a template mechanism, and a chemical network that produces a kind of primitive membrane that envelops the previous two (Gánti 1975, 1977, 1980). Computer simulations show that the chemoton system is highly stable, and it is able to grow and reproduce (Csendes 1984). The theory behind the chemoton model provides convincing evidence that all three subsystems might form on primeval Earth.

As was discussed, the evolution of present biological systems from one unique system of compounds, although theoretically not impossible, is very far from being proved. At our present state of knowledge it is more reasonable to assume that various chemical networks have evolved independently, and at a later stage they have begun to *cooperate* and unite (as described in the chemoton model).

Eigen (1971) studied a more developed stage of molecular evolution. He started off assuming that a molecule capable of replication already existed and analyzed the effect of mutation and competitive selection on its structure. The impressive mathematical apparatus used made this investigation well known and respected. Unfortunately, the author never has dealt with the replicating molecule's *origin* (see one of the modest criticisms by Ferracin, 1981). Recognition of the necessity of cooperation among the closed, cyclic replicative processes (Eigen and Schuster 1977, 1978a, 1978b) is a valuable by-product of this work. Eigen and Schuster call these cooperative replicative cycles hypercycles and consider their emergence as a consequence of competitive selection and the special hyperbolic kinetics of the replicative processes. (For the latest developments in hypercycle theory, see King 1981 and Müller-Herold 1983.) The hypercycle is based on a special mathematics and strongly differs from our *supercycle*, which is an organization of coreplicating compartments.

In my opinion, we need to explain the origin of both the topologically connected *compartments* and the organizationally connected *supercycles*. Replicative selection has played an important role in the formation of compartments and supercycles. In the phase of early evolution when replicative information was already present and replicative organization started to act as a selective force, functions of an organizationally higher type emerged. So far function has been defined in connection with components as an action by which a system's *component* is able to influence the probability of genesis or survival of another component. However, as a consequence of the above-mentioned nested nature of the biological system and its components, not only components within each individual RCN subsystem influence the probability of one another's genesis, but organization of a higher level emerges and RCNS, as components of the zero-system, begin to influence the probability of one another's genesis. This new interaction causes the appearance of functions through which an individual RCN is able to influence others in the interest of its own survival. These new functions belong to whole subsystems (RCNS) and not to separate components. Thus the description of these new functions must occur on two different levels, and the emergence of the higher level is an addition to the already existing molecular and "system" levels of the RCNS. This higher level of organization is expressed as a connection among various RCNS, and it directly causes the formation of the supercycles that are inseparable from the development of compartments of various natures.

To summarize: the information content of the whole zero-system grows during the nonidentical phase of replication because of replicative selection. Due to the emergence of new organizational levels, sets of components become diverse and separated topologically. These sets act as new unities and are called *compartments*. The whole process can be viewed as the *compartmentalization of replicative information*.

Responding to the effect of compartmentalization and Darwinian selection, the fidelity of replication continuously increases within compartments, i.e., in individual RCNS. Gradually, control mechanisms develop, and the fidelity of replication slowly approaches unity. This phase is reached at the end of a period of nonidentical replication with development of the genetic code, the mechanism of *identical replication*.

Formation of the genetic code and the phase of identical replication

Ever since the genetic code was understood and its universal importance recognized, evolution of the code system has been widely studied (Crick 1968, Orgel 1968). The two main questions—what was the mechanism of evolution of the code, and why did this type of code system happen to evolve?—are still open. Theoreticians propose the most diverse solutions to these questions. According to Woese (1967), the pivotal factor in the code system's development is the affinity between different amino acids and nucleotides. This assumption is vigorously debated, but recent data show that there is a strong correlation between the hydrophobicity of individual amino acids and that of the anticodons of their respective TRNA (Lacey and Mullins 1983).

Essentially the "dictionary extension" theory assumes that a smaller number of amino acids originally were coded, and incorporation of new amino acids into the protein-synthesizing machinery was made possible by the present-day three-letter code system's evolution (Crick 1967, Jukes 1967). The theory, known as "ambiguity reduction," appears to be the most firmly based. It contends that, initially, groups of amino acids were coded by respective groups of codons, and later under selective pressure a finer distinction within the groups developed (Fitch 1973). Eigen believes that the code system developed completely by chance (Eigen 1971).

Besides the development of the code system, great attention focuses on the evolution of the mechanism of protein synthesis, with special reference to the roles of TRNA and ribosomes (Crick et al. 1976, Batchinsky and Ratner 1976, Ratner and Batchinsky 1976, Mizutani and Ponnamperuma 1977). Among diverse theories, the most promising are based on experimental data that try to account for the physico-chemical connections between the structures of protein synthesis, i.e., proteins, amino acids, and nucleotides.

The accuracy of replication through the genetic code system's mechanisms reached a maximal level very early and has remained close to that level for billions of years (Kimura and Ohta 1971). In the case of replication of nucleic acids this means one replicative error in about every ten to every 100 million pairs of nucleotides (Drake 1974, Watson 1976). A more accurate replication would be a selective disadvantage (Maynard-Smith 1978).

After examining the genetic code in some detail, another very important mechanism—*recombination*—which assures the accuracy of replication by removing whole sets of inaccuracies from the genetic pool of a species can only be mentioned here.

The consequence of identical replication: convergence

As a result of identical replication, changes in structures that provide various selective advantages became irreversibly fixed in the system's components. In the phase of nonidentical replication, divergence was the most characteristic feature of evolution, as the individual components were highly variable and a great variety of new structures appeared. In the phase of identical replication, the functional synchronization of the supercyclical units continues within the compartments. Thus variability decreases, and a kind of *convergence of the control* systems evolves. The divergent and convergent phases of molecular evolution have been outlined by Kuhn (1976). As a result of convergence, the different replicating components become even more tightly connected; functions also become more coordinated; and, finally, the cooperating components form a new replicative unit. The new replicative unit identically reproduces its own structural and functional information content in each replicative cycle. This is the point where the compartment reaches its most stable state.

Parametric and functional information in proteins and gene duplication

Since the amino acid sequences of numerous proteins became known, efforts have been made to gain information concerning the mechanisms of evolution from the amino acid sequences of specific proteins of different species. The *functional information* content (loosely resembling our definition) of proteins is an accepted concept. Generally, the sections of the polypeptide chain containing amino acids crucial for a given protein's catalytic or other function are almost uniform in the different species. Namely, in these segments the number of "accepted" mutations is very low. At other parts of the chain a variety of substitutions might occur. Based on these data, a kind of "functional density" of a given protein can be calculated. This is very high, for example, in histones, and low in the case of immunoglobulins.

Zuckerkandl (1976a, 1976b) makes a distinction between "functional" and "general" information; namely, general information is carried by amino acids that are not directly necessary for the specific function of the protein but account for its "general" properties such as solubility, surface charge, etc.

There is an obvious analogy here with my notions of functional and parametric information introduced earlier. Holmquist and Moise (1975) compared the amino acid composition of a series of proteins to a random composition calculated from the code dictionary. They found that the real amino acid content of proteins differed from a random distribution; moreover, the difference was nearly the same within each protein family. They attributed these deviations to the information necessary for function. It seems that this also is an example of the dichotomy of parametric and functional informations. In proteins, therefore, parametric information can be calculated from the code dictionary. A study of the protein cytochrome-C proteins isolated from presently existing species has shown that the density of functional information increases and at the same time "evolutionary noise" decreases in the course of evolution. The "expression" of cytochrome-C becomes simpler in higher organisms (Reichert et al. 1976). These results support the concepts of parametric and functional informations, which we introduced earlier.

With the development of the genetic code, the parametric information content of nucleic acid molecules changed decisively. It can be calculated best by comparing it to a uniform nucleotide distribution based on the code dictionary. Any deviation from this distribution indicates the appearance of functional information.

The phenomenon of the "evolutionary clock," widely discussed in the literature, as well as the related theory of neutral mutations should be treated here, even if only briefly. On the basis of biochemical data it appears that, considering a longer period of several hundred million years, the mutations of DNA of various species occur at a constant frequency, independently of the speed of evolution of the relevant species assessed from morphology or other phenotypical characteristics. For example, frogs have retained their original morphological features for about 150 million years; still, their various proteins change at the same rate as those of mammals, which have evolved only in the last 80 million years and are in a period of extremely fast evolution (Fitch 1973, Wilson et al. 1977). According to the neutral mutation hypothesis, this is due mainly to the indifference of most of these mutations from the standpoint of an organism's survival (Kimura 1968, 1969, 1977).

Here it is important to point out that the neutral hypothesis of mutations and the implications of parametric information concerning the protein structure are not contradictory. The parts of the structure containing parametric information do not contain functional information; consequently, they are hardly subject to selective influences. Their mutations might be caused by the random fluctuations of parameters.

Also there is no contradiction between the hypothesis of neutral selection and that of competitive selection, if we assume that the latter is only one kind of possible selection type. As we have discussed, we must consider at least two selection types. One of them, beyond doubt, is competitive selection. The other is replicative selection, which is a more general phenomenon because it removes all changes from the system that do not fit into or inhibit the replicative mechanisms. Since it removes only the negative changes, it permits all others not harming replication; that is, it permits neutral mutations. Replicative selection continuously increases the choice of components. It seems that competitive selection is a tuning mechanism, which makes finer adjustments, compares the already fixed variants, and under restrictive conditions (restricted space or resources) allows the fittest to survive. We will return to this question in the discussion of problems of macroevolution.

The relative steadiness of the "evolutionary clock" could result from the ultimate genomic fixation of every evolutionary change at levels above the genome, i.e., the cell, organism, or ecosystem level. Accordingly, the nucleotide sequence of DNA contains information concerning not only the functions of proteins inside the cell, but the corresponding structures and behavior of higher levels. Even when evolution has finished at a lower level, further changes on higher levels keep moving the "evolutionary clock" until the whole terrestrial system reaches an evolutionary equilibrium. Then the "evolutionary clock" will probably stop.

Concerning the question of the general trend toward increasing replicative information, some stimulating ideas are provided by the theory of gene duplication. For a long time it has been argued that the exchange, deletion, or addition of nucleotides as the only source of mutation can completely account for evolutionary variability. Today we know that it cannot. In addition to point mutations, several mechanisms are known to be essential sources of variability. The most important among them is gene duplication. It is assumed that during evolution some genes, eventually a complete genome, may duplicate due to some mutagenic effect. One copy performs its original function, while the other, released from the selective pressures of competitive selection, goes through a series of mutations and gains new functions, thus enriching the adaptive possibilities of the organism (Ohno 1970).

By examining the amino acid sequences of proteins it is possible to draw conclusions regarding gene duplication. This has been shown in regard to the most diverse proteins (Woese 1971, Reek et al. 1978). Computer analysis of the amino acid sequences of proteins appears to prove the role of gene duplication (Barker et al. 1978). Complete genome duplication has been demonstrated in prokaryotes, namely, in Mycoplasms (Wallace and Morowitz 1973), in *E. coli* (Zipkas and Riley 1975), and in eukaryotes in which this phenomenon has been known for about seventy years (Ohno 1970, Markert et al. 1975). Gene duplication is actually a very interesting form of the increase of replicative information. The only special event during gene duplication is that the replicates of the genes produced in the replicative cycle do not separate but remain together in the cell. Their further replication continues undisturbed since both have equal replicative information content. The information in one of the replicates remains unchanged; the other one changes the specificity of its functions through a series of mutations but retains its general replicative function (replicative selection in action!)—and so the transformation of a specific function is made possible. For the cell as a replicative unit, gene duplication means an increase in the amount of information-containing structures. The subsequent series of mutations leading to the new function decreases redundant replicative information that arose due to gene duplication. The new functional information replacing the redundancy further increases the cell's complexity. This phenomenon completely agrees with the notion that replicative information is continually increasing through evolution. Potentially, each replicative cycle—i.e., each cell division—can be a starting point of an Ohno process. The duplication of the cell's genome is only a first step in the development of a more complex structure. This can be followed by a division, so that there is no second step, or a second step in which the duplicated genome remains partially or completely together. The growth of functional complexity requires further mutations. It is obvious that the above process requires a closed space, a trend of compartmentalization, and convergence. In a spatially open system convergence cannot develop; thus compartmentalization, being the result of forces manifested in the functions of molecules, provides the conditions for the development of convergence.

A most interesting phenomenon, which cannot be explained by competitive selection, was recently discovered. Certain sequences of DNA were found to occur in numerous identical copies scattered in the whole genome. Two kinds of such "anomaly" have been found. The so-called "short repetitions" consist of less than five hundred nucleotides and occur in 10,000 to 100,000 copies per genome. The "long repetitions" may be longer by an order of magnitude, and the number of copies is about the same as in the short repetitions. Ever since the discovery of this curious feature, its biological function has been sought, especially as the genome of the higher animals contains repeated sequences in 5 to 30 percent of the total genome. However, even the most thorough research for phenotypic expressions of these repetitious DNA sequences has been unsuccessful. Theoretical research also has started, and many brilliant speculations have been published (Orgel and Crick 1980, Doolittle and Sapienza 1980, Dover and Flavell 1982). The most fascinating one seems to be the idea of functionless, "parasitic" DNA. It is difficult to explain why competitive selection has not removed these functionless sequences. The economic argument that it costs the cell nothing to produce these pieces is not very convincing. Some researchers consider these segments to be the structural bases of new enzymes or other proteins that may evolve in the *future* (Ohno 1984). Although this is not impossible, the idea seems to contradict Darwinian selection theory, since it suggests the fixation of a character that carries only a future use for the organism.

The presence of parasitic DNA can be explained on the basis of replicative selection theory. All those changes that do not harm the replicative process may be incorporated—inserted into the genome. The emergence of parasitic DNA segments is just this type of change. Both replicative selection and competitive selection mostly affect the structures necessary for replication because of their very function. Structures replicating accurately, but having no function, do not disturb the process of replication as long as their proportion in the genome remains under a certain limit. An organism with 90 percent functionless DNA probably could not survive because if the absolute size of the genome is kept under certain limits, there would not be enough functional DNA.

The repetitive DNA sequences are in a certain sense carriers of the "clearest" replicative information because it has no other function but to participate in the replication.

The autogenesis of the cell

The cell is the final, most organized product of the first level of evolution. The spontaneous *autogenetic* process that has led to the cell can be summarized as follows:

In the thermodynamically open zero-system of primeval Earth material flows excited by the energy flows of the system begin and spontaneously form simple autocatalytic reproductive chemical networks (RCNS). RCNS are very effective *organizing* agents; they are precursors of more developed systems. Production of compounds becomes controlled; protofunctions of the compounds evolve to functions; and the compounds evolve to components. Simple chemical reproduction evolves to *replication*. As the fidelity of replication grows, functional connections among components become increasingly "tuned," and *compartments* (sets of components that replicate together) are formed. The replicative information content of the whole system continuously increases.

In the replicative network formed in the final product—the cell—DNA carries a description of the specific constraints necessary for replication. Most of the replicative information is concentrated in DNA. All other parts of the cell may be viewed as a constructor that interprets the replicative information and constructs the whole replicative network.

The size of the compartments is determined by the forces of molecular interactions, i.e., the constraints on the molecules. The critical distance is not longer than a few centimeters, the distance from which two molecules can influence the probability of each other's genesis. The final compartments of molecular evolution, the cells, may not be larger than the size determined by molecular forces; actually, most of them are smaller. Since the volume of the cell compartment is several orders of magnitude smaller than the volume of the whole zero-system on the surface of Earth, a great number of *different* cells could evolve in the early zero-system.

The cell, as a "system," represents an almost closed network of components and component-producing processes, which in turn continually produce the same network. For *maintenance* of the cell, several thousand kinds of molecules have to be produced continuously in a concerted process, and production of the same set of molecules is also necessary for reproduction of the cell. Therefore, the cell's maintenance and *reproduction* are based on *replication* of molecular components in time and space.

The cell's evolution, having reached the final phase of convergence, has been completed. For example, *E. coli* has developed in the intestines of mammals and is known to have existed in an unchanged form for some 10 million years; that is, its inner structure with its unchanged quality has a durability comparable to many geological formations. The essence of the replicative information's convergence is this extreme *inner stability*. It is no contradiction that a cell may change in any phase of convergence for external reasons. A consequence of the inner stability of a cell is that as its "external" stability decreases, its existence becomes more and more dependent on the constancy of external parameters.

We must note here that the above description of the cell's *final* state largely agrees with the autopoietic model (Varela et al. 1974, Zeleny 1977). However, the authors of that model have neither used the concept of information nor analyzed the question of the *origin* of autopoietic systems, which is the very subject of our study. Autopoiesis is the model of cell maintenance, and in its present form it is unsuitable for analyzing evolutionary processes. Our theory of autogenesis, however, serves as a model for studying the evolutionary development of cells and other systems (Csányi and Kampis 1985).

The autogenesis of eukaryotes

Conditions for life following the emergence of replicating cells

Until recent decades the evolution of cells was a neglected area, mainly because the search for fossil remnants of cells had been unsuccessful and also because biochemical and molecular-biological methods suitable for finding traces of the evolutionary process in contemporary cells were only recently developed.

For a long time biologists were convinced that two basic types of cells existed: *prokaryotes*, i.e., cells without nuclei, which were considered a more primitive type; and the larger, more complex *eukaryotes* having special membranes that enclose the distinct nuclei and other cell organs.

The prokaryotes are small cells (several micrometers in diameter) shaped like spheres or rods and covered with a membrane envelope and a cell wall. DNA and ribosomes, which are the sites of protein synthesis, are swimming freely in the cytoplasm. The eukaryotes are larger cells, sometimes discernible to the naked eye. DNA is distributed in chromosomes that also contain some proteins and RNA. The chromosomes are located in the cell nucleus, which is separated from the cytoplasm by a nuclear membrane. The eukaryote cells may contain some of the various cell organelles, e.g., mitochondria that are the organs of cell respiration, or in plant cells, chloroplasts, the cell's organ for photosynthesis. The ribosomes of the eukaryotes are bound to a special membrane, the endoplasmic reticulum, and many eukaryotic cells have Golgi organs that are responsible for excretion. The baglike lysosomes are digesting organs, and the centrioles move and arrange the chromosomes during cell division. Generally, even the simplest eukaryote is more complex than any prokaryote.

There is a further addition to these basic morphological differences: prokaryotes multiply simply by the cell's division, while in eukaryotes, besides cell division, sexual forms of propagation have evolved (although there are some similar processes, such as conjugation, transformation, or transduction in prokaryotes, which also can be considered the most primitive forms of sexuality).

The eukaryotic cells are capable—at least in certain phases of their life—of going through *meiotic cell division* to form sexual cells. The sexual cells contain only one-half of the complete set of the normal cell's chromosomes, but they contain them in new recombinant forms. By unification of the sexual cells a *zygote* is produced, which may propagate by mitotic cell divisions during subsequent cell cycles. Meiosis is the most advanced form of genetic recombination, and it probably speeded up the rate of evolutionary changes.

Let us now survey the living conditions on Earth from the time of the first evolving molecular systems' appearance to the emergence of eukaryotes. According to available data, the atmosphere of primeval Earth was probably strongly reducing (although there are other theories, and the question is still open); it contained carbon monoxide, ammonia, water, cyanide, elementary hydrogen, nitrogen, and only traces of oxygen. The atmosphere did not contain much oxygen until 1.5 billion years following life's origin. It also is supposed that on the Earth's surface various simple organic compounds were abundant and the first progenotes were heterotrophic, i.e., they were feeding on abiogenic organic compounds (figure 11).

Figure 11 Time schedule of early evolution.

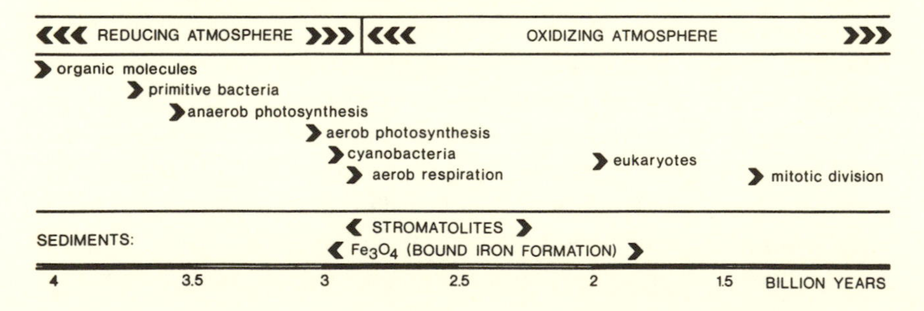

According to various studies, it seems probable that ammonia was the first major component of the primitive atmosphere that depleted, and surviving progenotes were able to develop chemical processes and enzymes needed for the fixation of free nitrogen. Possibly present-day nitrogen-fixing bacteria, living symbiotically with higher plants, are the closest descendants of these early progenotes. These modern bacteria can fix nitrogen only in a reducing milieu, which also supports the idea of the early atmosphere's reducing nature.

About the same time other cells emerged that were capable of the direct use of solar energy through *photosynthesis*. Their relatives have been living since then. There are many differences between bacterial and vegetal photosynthesis. The most important difference was probably the oxygen intolerance of early bacteria. The species still living in a reducing environment and performing photosynthesis accompanied by oxygen production are probably descendants of these early species. Oxygen-producing photosynthesis was much more efficient than nonoxygenic photosynthesis, and it certainly enhanced the rapid propagation of the species having this advantage. This resulted in an enormous "*pollution*," i.e., the accumulation of free oxygen. Oxygen was very toxic for all living beings, even for the species producing it. A fast selection process began, and species were able either to evolve resistance to oxygen by transforming their metabolism, or they were pushed into niches such as marshes and other environments poor in oxygen. The present-day cyanobacteria descend from the early prokaryotes that transformed and tolerated oxygen. Cyanobacteria appeared in abundance some 2.3 billion years ago, but fossils as old as 3 billion years also are known. Their traces can be found in many forms, but mostly in fossilized bacterial mats called stromatolites. After their spread on the surface of Earth the formation of an oxidized inorganic compound (Fe_20_3) started, probably because of free oxygen in the atmosphere. Free oxygen brought about another change. In the lower layers of the troposphere, ozone appeared, which very effectively filtered the ultraviolet component of radiation coming from the Sun. This in turn might have decreased mutation rates and served as a protective shield for living beings. These changes transformed the whole living world, as millions of species disappeared, making room for the more developed oxygen-tolerating ones.

Present cyanobacteria prefer an atmosphere with 10 percent oxygen, so our atmosphere with more than double that concentration must be the by-product of the photosynthesis of higher plants.

Because of their more effective photosynthesis, cells containing chloroplasts displaced cyanobacteria. Also with the advent of photosynthesis organisms appeared that, although not able to perform photosynthesis, could tolerate free oxygen and in fact required it for an effective metabolism. These species are the modern heterotrophic *aerobic* organisms.

With some uncertainties a timetable of the most important changes can be estimated. Chemical evolution was quite fast; it may have taken only a few million years, approximately 3.8–3.9 billion years ago. Later, 2.3 billion years ago, there was a considerable amount of free oxygen in the atmosphere, and highly developed eukaryotic cells appeared approximately 1.5 billion years ago (Schopf 1978).

Selection and symbiosis in cellular evolution

Explaining the evolution of complex molecular or other mechanisms simply on the basis of competitive selection always proved difficult for neo-Darwinian theory. For example, organelles of the eukaryotic cells are very complex and of effective "design," and it is very hard to suppose their gradual appearance. In general, the formation of cell organelles was explained by the compartmentalization of DNA (Raff and Mahler 1972). DNA can be found in both the ATP-producing mitochondria and the photosynthesizing chloroplasts. These include genetic information for several proteins needed for the work of the respective organelle. It seems logical to suppose that the necessary genes, mitochondrial oxidation, for example, were separated from the bulk of the DNA and localized close to the processes they control. Logical or not, however, it is wrong. In the mitochondria besides DNA there is protein synthesis; it was shown that in some very important details this synthesis much more closely resembles similar processes in prokaryotes than those in eukaryotes. Following these important findings, biologists started to search for other explanations. It is interesting that a merely speculative hypothesis was published by J. E. Walkon as early as 1920, stating that mitochondria were originally bacteria; however, nobody at the time took him seriously.

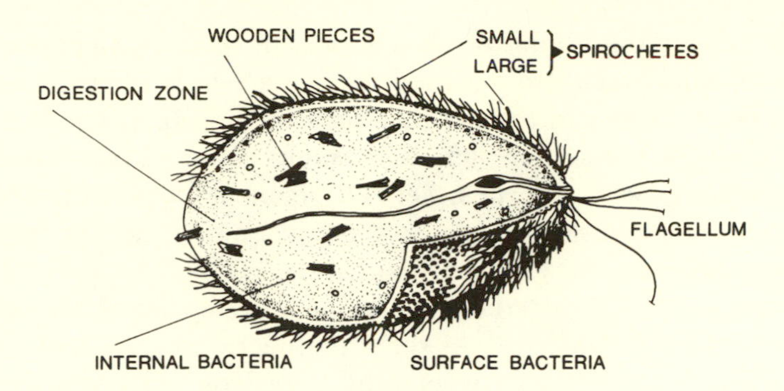

Figure 12 *Myxotricha paradoxa* (after Margulis 1971).

There are many examples of symbiosis of unicellular organisms (Margulis 1970). *Paramecium bursaria,* a eukaryotic protozoon, for example, upon meeting a green algae (Chlorella) cell swallows it without further digestion. The cell of Chlorella remains intact and swims around the cytoplasm of the Paramecium and even reproduces there. In the dark the Paramecium feeds its Chlorellas, but in the light the algae cells intensively photosynthesize, and parts of the product are consumed by the Paramecium. This is a fine example of the mutual benefits of symbiosis. It also was observed that once the Paramecium has Chlorellas, upon meeting any further Chlorella cells it not only swallows but also digests the newcomers, so the Paramecium recognizes its "own" algae cells.

Another interesting example of symbiosis is the life of *Myxotricha paradoxa,* also a protozoon, living in the guts of certain Australian termites (figure 12). This big protozoon lives symbiotically with three different bacterial species at the same time, and its relation to the termites also is symbiotic. Attached to the surface of the Myxotricha live spirochetes, which help the bigger protozoon swim fast by beats of their flagella. Each spirochete is in a symbiotic relationship with another bacterial species located on the surface of the Myxotricha. Last, a third species of bacteria lives in the cytoplasm of the protozoon and contributes to its digestion with its enzymes, breaking down cellulose. The termites cannot digest cellulose; without Myxotricha they would starve to death.

These examples show that symbiotic relationships are not infrequent among unicellular organisms, and in some cases prokaryotic and eukaryotic cells are involved in very effective interactions. We have every reason to assume that such relationships also occurred in the earlier phase of evolution. Margulis assumed that mitochondria, chloroplasts, and the other cell organelles were evolved from symbiotic relationships similar to those just discussed. The theory is supported by various experimental results.

From analysis of bacterial proteins and RNAs the following sequence of events can be inferred. As the atmospheric oxygen content was changing, anaerobic species of cells were largely pushed out, but the remaining ones survived in ancient marshes, which provided a restricted but oxygen-free niche. The first symbiosis had occurred between an amoebalike anaerobic *urkaryote* (the ancestor of eukaryotes) and an aerobic prokaryote, already able to use oxygen for producing ATP. The advantages of this connection are clear. Symbiosis of the two resulted in organisms able to live in both aerob and anaerob niches. Present-day eukaryotes use oxygen to burn carbohydrates, and these processes take place exclusively in the mitochondria. The mitochondria and most probably its prokaryote ancestors cannot respire without free oxygen. In anaerobic conditions an alternative anaerobic metabolic pathway of energy production, fermentation, is used, employing specific cytoplasmic enzymes. Although fermentation provides much less energy compared to mitochondrial oxidation, it allows the cell to survive for a considerable time. In a subsequent symbiosis a spirochete joined the amoeba, which already had the advantage of aerobic respiration. Amoebas themselves were very slow, but the spirochetes, helped by flagella, could swim around with great speed. Centrioles, the molecular moving apparatus supposedly originating from spirochetes, provide (besides fast locomotion) help in the rearrangement of chromosomes during mitosis—the reproduction process of the eukaryotic cell. From these mobile cells *fungi* and *animals* evolved. A different type of symbiosis—between a photosynthesizing prokaryote and an amoeba—led to the evolution of chloroplast from the former (figure 13), creating the ancestors of modern *plants*.

According to a now widely accepted theory, during the evolution of these symbionts the participating cells lost their independence and except for a small fraction the DNA of their genome also melted into the common eukaryote genome (Schwartz and Dayhoff 1978).

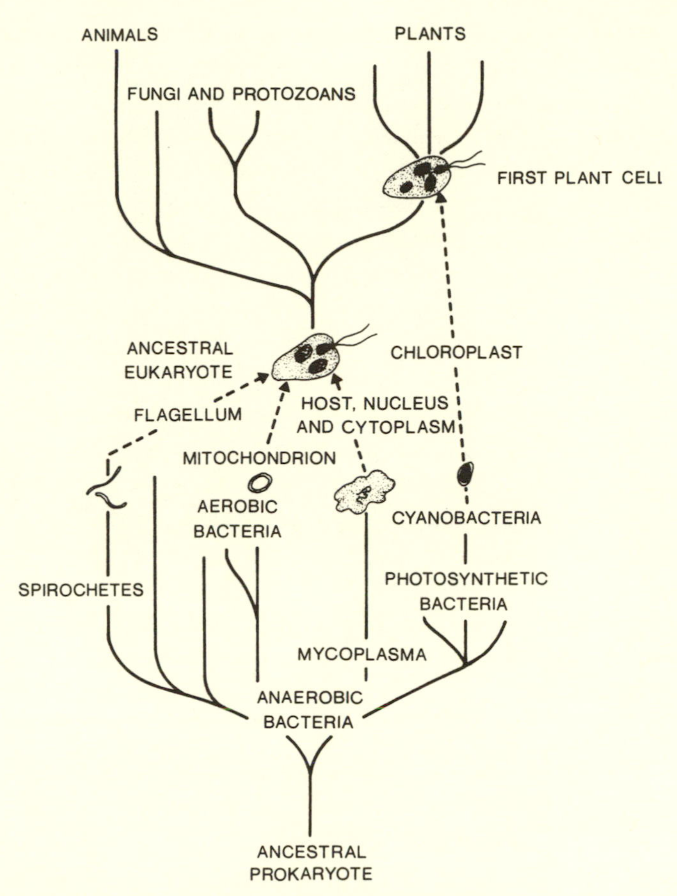

Figure 13 Origin of the eukaryotic cell (after Margulis 1971).

Relationships among the unicellular organisms

Analysis of the evolutionary relationships of present-day unicellular organisms (such as bacteria, algae, and protozoons) and their higher descendants (such as animals, plants, and fungi) was greatly promoted by Margulis's symbiosis theory. Advanced molecular biology provided the means for these studies. The fine structures of proteins, RNAS, and ribosomes were analyzed and compared, and evolutionary relationships could be outlined from the data obtained. (The evolutionary tree of contemporary cells is shown in figure 14.) Recent data show that from progenotes, the most ancient cells, four main

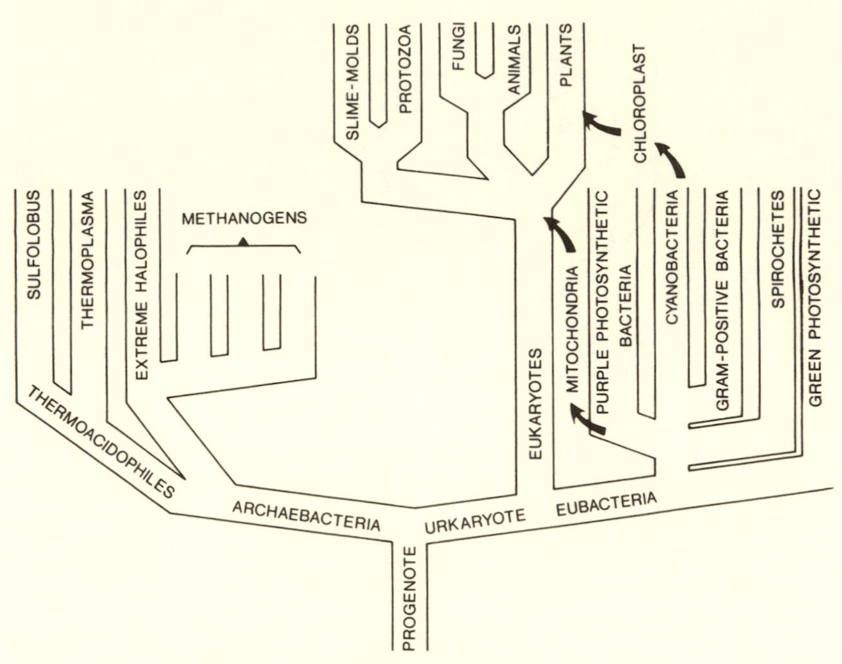

Figure 14 Origin of living organisms (after Woese 1981).

branches evolved. There are two branches of *archaebacteria*, the sulfur bacteria and another including methanogenic and halophilic bacteria. The methanogenic archaebacteria are anaerobic; the halophiles and the sulfur archaebacteria are aerobic. The third branch includes *eubacteria*—all the contemporary prokaryotic species except archaebacteria. Urkaryotes belong to a fourth branch. Plants originate from the symbiosis of urkaryotes, cyanobacteria, and the ancestor of the mitochondria. Animals and fungi originate from a branch without chloroplast (Woese 1981, Lake et al. 1984). Another valuable finding is that a constituent of the eukaryote ribosomes has originated from the archaebacteria, confirming that all main branches of ancient evolution contributed to the evolution of the highest developed eukaryotes (Chamblis et al. 1980).

Prokaryote multicellulars?

In my discussion of eukaryote evolution one important question has been avoided, namely, whether or not *multicellular prokaryotes* have evolved? The fossils are not decisive enough, but there are some interesting findings that indicate the past existence of multicellular prokaryote organisms (Nagy 1974). These fossils are about 2.3 billion years old; i.e., they originate from a period before the emergence of eukaryotes. The possible existence of multicellular prokaryotes is perhaps even more convincing when we think of the various forms of symbiosis as a special way of forming a multicellular organism. With some generosity, Myxotricha can be considered a multicellular being.

Therefore, we may put forward a hypothesis of a two-phase evolution of multicellular organisms, assuming that the eukaryotic cells are nothing other than highly organized descendants of ancestral multicellular prokaryotic organisms. Multicellular eukaryotes may represent an even higher level of organization in which the eukaryotic cell is only a component of a higher system.

We may now examine the evolution of the cell within the theoretical framework of an autogenetic model.

As soon as genetic mechanisms producing identical replication appeared at the level of molecular evolution, convergence and compartmentalization would have created a perfect replicating system: the *cell*. The first cell species, most probably various anaerob progenotes, have spread across the entire surface of Earth, ranging from ice deserts to hot springs, and their descendants still live in these varied niches. We may confidently suppose that *millions* of species were created through environmental variability and random fluctuations.

This great variety of organisms created a *new evolutionary zero-system*. The anaerobic cells, the *components* of the new zero-system, formed and decayed continuously in great variety. Accurate formation of these components was ensured by identical replication, while variability was generated by mutations. These components were set in motion by the energy flow of the system and participated in the Earth's great material cycles. Because of incessant competition for material and energy sources, *functions* emerged through their mutual interactions. Cells sharing a common environment began to influence the probable genesis of one another by compounds secreted and adopted.

What are the system precursors of this new zero-system? Interactions of cells represent functions, while a given network of these functions represents a *multicellular organization*. The minimal condition for such multiple cell organizations to exist is their replicative nature. This minimal condition promotes the action of both replicative and competitive selection mechanisms, as was described for the molecular level. An important difference, however, is that the selective effects of organization at the cellular level, affecting cell interactions, also affect the interactions within the cells at the molecular level of organization. Conditions at the higher organizational level act as selective forces producing specific changes of the molecular components; consequently, the hierarchical cycle of causality that characterizes all living systems is formed. Structures at the higher level of organization influence the selective processes at the molecular level, but the resulting molecular changes also are manifested at the higher level. This interaction leads to a positive feedback, which in the long run separates the internal structures of the higher-level organization from processes at the lower levels, making them almost independent.

We now have discussed all system characteristics that are necessary for application of the autogenetic model. We believe, moreover, that at present only the theory of autogenesis can provide an explanation for this phase of evolution.

We must define some additional parameters of the autogenetic system model. In the cellular zero-system, cells are continuously produced. At a hypothetical initial moment, when the cells do not yet interact, i.e., when they are without functions, the types and numbers of cells exactly reflect the parameters (temperature, concentration of food, materials, etc.) of the new zero-system. At this moment the distribution of cell components reflects the *parametrical information* content of the cellular zero-system. As soon as the cells start to interact, *functional information* emerges; i.e., certain kinds of cells will grow and propagate successfully, while others become extinct or sparse. *Replicative information* at the cellular level is a description of the specific constraints accounting for the interactions of cells, but it does not obey the chemical laws valid on the lower molecular level. Moreover, as all the information is carried by the system's molecular components, the amount of replicative information—and the stability of the system—is increasing at the expense of parametrical information. It also is evident that the appearance of functions is not an all-or-nothing phenomenon; it is gradual and rebounds to the previous phase of evolution. Nevertheless, it is worthwhile to separate the various forms of information because this allows us to recognize analogous phenomena at the various levels of organization.

Perhaps it is unnecessary to demonstrate separately that in the early phase of multicellular evolution *autocatalytic* interactions of the cells quickly emerge. If cell A increases the probability of genesis of cell B, and cell B through some metabolite does the same to cell A, then in this autocatalytic cycle the two cells mutually enhance each other's rates of replication and gain a considerable advantage over other cells. Complexes of cells connected by mutually advantageous functions can easily incorporate other kinds of cells, and thus stable *cell ecosystems* may arise (Margulis 1970) that correspond to multicellular *supercycles*.

Multicellular evolution also starts with nonidentical replication. In this phase there already are some functional connections among various cells, and the quasi-ecosystems reproduce better, but the replicative mechanisms of the partners are still separate and individual, as was shown above in the discussion on instances of symbiosis. In cellular ecosystems functional differentiation emerges as a necessity; higher supercycles are formed; and *convergence* of replicative information on the multicellular level starts. Selection favors cell ecosystems that are more successful in coordinating their replication than others. The fidelity of replication of cell ecosystems increases surely, and, as a consequence, *compartmentalization* begins. Although we do not have enough data about phases of this process, it might be stated with some confidence that the autogenesis of associations of prokaryotes reached the phase of identical replication by this time. Its mechanism is mitosis, and its final product is the *eukaryotic cell*.

Modern eukaryotic cells rarely reach the size of one millimeter, but there are many fossils showing that in the early phase of eukaryote evolution giant (over one centimeter in diameter) unicellular organisms were common (Schopf 1978). These giants perhaps had more complex cell organelles, but they were pushed out by the better-developed multicellular eukaryotes.

3 The Evolution of Multicellular Organisms

The first period of the evolution of prokaryotes

Multicellular organisms appeared about 750 million years ago, relatively late in the course of evolution (Maynard-Smith 1969). It would be a simplification to assume that autogenesis of multicellular organisms is nothing more than a repetition of the organizational events that had led to the appearance of eukaryotes. This would mean that formation of this new type of cell created a new zero-system with eukaryotic cells as components, and that interactions among them would lead to multicellular organisms. Some remarkable phenomena suggest consideration of other possible scenarios.

No separate mechanism in multicellular organisms for reproduction resembles the cell division of the prokaryotes or the mitosis of unicellular eukaryotes. Reproduction processes of the latter two are essentially simple divisions, with a roughly equal distribution of components between the two daughter cells. Reproduction of multicellular organisms by a similar process is very rare; it occurs in lower species, but even in those the analogy can be debated, as it is simply a manifestation of the ability of the lower multicellular organisms to regenerate. Division and regeneration of parts of these organisms are only the exploitation of the reproductive mechanisms of these components, the cells, and there is no special new way of reproduction that would be characteristic of the new organizational level.

The regenerating ability of higher organisms is minimal, and their reproduction is characterized by a transition through a *cellular phase*. We may safely state that the higher multicellular organisms are not able to properly replicate themselves. Instead of replication, their structural organization is disassembled in each generation, and through the zygote formed from germinal cells an *autogenetic process* called *ontogenesis* re-creates the multicellular organism.

Another important point is that the evolutionary mechanisms that have led to the emergence of multicellular organisms also must explain the formation of *organs* as compartments. The difficulty is very similar to the one encountered in explaining the emergence of cell organelles; i.e., it has to be assumed that the eukaryote genome somehow becomes compartmentalized, and the organs are formed later by differentiation. Although such a model cannot be rejected, we first put forward a new one that fits into the framework of the autogenetic theory, taking the risk of appalling some of our biologist colleagues.

A model of the origin of multicellular organisms from prokaryotes

We should turn back for a moment to the autogenesis of eukaryotes. We may assume that many different prokaryote ecosystems have existed. *Cooperation* of participating cells has started the process of integration, and in present eukaryotes very complex cell organelles can be found. Based on this analogy, it seems reasonable to suppose that formation of the organs of multicellular organisms also has resulted from the functional cooperation of various ancient prokaryotes.

Let us suppose that the various organs or at least the basic types can be traced to prokaryotes. We take for an example an ecosystem of prokaryotes with one member species secreting detoxifying enzymes that help remove harmful substances from the environment. This function is clearly advantageous for the ecosystem's other members, and it is similar to the animal liver's function. We then assume that an evolutionary process of *integration* began, just as in the case of cell organelles, and that the cell species with detoxifying enzymes has been incorporated into the system of prokaryote cell species. The only remnant of its early autonomy is its existence as a compartment, i.e., the liver organ within the organisms. The eukaryote genome is enormous in size, including most of the genomes of ancient prokaryotes that have formed the cell organelles. We have no reason to suppose any special obstacles to the incorporation of genomes of prokaryotes, which later could have formed nervous tissue, liver, kidney, etc.

This line of thought is supported, at least logically, by the following argument. If we assume that the formation of a multicellular organizational level was preceded by the eukaryotic cell's evolution, this implies that multicellular organisms have originated from *one or a few* eukaryote species. Thus the primary ancient eukaryotic cell would cease to be autonomous; several daughter cells would remain connected; and mutational events would accumulate during a long evolutionary period in which the completely similar cells became *differentiated*.

This concept can be debated on two fronts. First, it is known that competitive selection is strongest among similar organisms because they need the same resources. Thus it is more likely that *different* kinds of cells start to interact and tolerate their mutual presence in space because they then can share some resources and benefit from the advantages of cooperation, even if these initial advantages are not very great.

The other argument against multicellular evolution from a single eukaryote species is that simultaneous, successful selection for all necessary properties of a highly complex multicellular organism is rather unlikely. When selective pressure is directed toward multiple properties, the incidence of favorable mutations is almost negligible. It is much more likely that selection of multicellular organisms began when ecosystems including cooperative prokaryotes evolved. In this case selection had to work only on the refinement of a primitive organization, which may have been much easier and faster.

Similar thinking was used in the famous watchmaker analogy by Simon (1962) illustrating the origin of complexity. The parable is about two craftsmen assembling watches. Each watch contains 10,000 parts. Work is occasionally disrupted by phone calls. The first master organizes his work so that he makes subunits, each from 100 pieces, and later he assembles a watch from 100 subunits. The other master assembles the watch piece by piece. The phone rings frequently, and the average time spent between two calls is enough to assemble 150 parts. If he suspends work, the already assembled parts fall apart; only the first master's subunits or whole watches are stable structures. The first master finished his first watch after eleven calls, but the other master will never finish even one because his incomplete assemblies fall apart after each call.

The parable's essence is clear and is valid for the hypothesis of the eukaryotic cell's evolution. The complex cell organelles of eukaryotic cells have not been formed by differentiation of the genome but from subunits consisting of cooperative prokaryotes. It is highly reasonable to suppose that a similar mechanism also has been at work in the evolution of multicellular organisms.

The logic of the separation of the evolution of eukaryotes from the evolution of multicellular organisms can be questioned on other grounds as well. It is possible, for example, that the appearance of the autonomous eukaryotic cell is only a by-product of multicellular evolution. All modern multicellular organisms are eukaryotes, and the free-living unicellular eukaryotes may well represent a dead end in evolution, or they may be the result of a devolutionary process.

A model of multicellular evolution from eukaryotes

In addition to this model for deriving multicellular organisms from prokaryotes, we will consider a classical model of eukaryotic evolution. Most biologists accept the idea of the evolution of higher organisms from early free-living unicellular eukaryotes. These theories suggest that multicellular evolution began independently several times. As a first step, a "colony" of an ancient eukaryote's daughter cells would form, and a long evolutionary differentiation of the genetically identical cells would result in the various organs and tissues of multicellular organisms.

We prefer an explanation by the autogenetic model based on *cooperation of different cell species*. Cooperation is the most important factor in the origin of functions. Different mutually interacting cells create a network of functions in the form of an ecosystem of the eukaryotic cells. The cooperating community of cells is a new *system precursor*, and its replicative organization entails replicative and competitive selection. A new level of organization is formed, and its replicative information begins to increase through convergence and compartmentalization. At the end of this process integrated *multicellular organisms* emerge, and the creation of the *organizational level of organisms* is completed.

Cooperation of identical cells would face many obstacles. First of all, there would be competition. Cooperation of different cells is more plausible. We can regard cooperation as a kind of "trap," for little advantages can "lure" individual cells into it, and later the very advantage of cooperation would "compel" participants to differentiate and thus further increase their benefit from cooperation. "*Trap of cooperation*" is certainly an important force in evolution, and its validity is well established. It would be interesting to reexamine our knowledge of ontogenesis and multicellular evolution from this point of view.

Two facts that seem to support this new approach will be mentioned. The first is a common observation concerning cells taken from various tissues of different animals and mixed in a suitable medium. They form aggregated groups—which interestingly are *tissue-specific* —independently of the species of origin. Thus brain cells aggregate with brain cells of other species, and kidney cells with other kidney cells, no matter what cells of whatever species have been mixed. Therefore, the type of tissue is a very ancient and common property. The other fact concerns various worms. During ontogenesis of Ascaris and Sciaria species, somatic cell lines form that contain different sets of chromosomes (i.e., the tissue cells are *genetically* different), and they obtain only certain sets of genes from the omnipotent zygote (Herskowitz 1973). Various interpretations could be found to explain this phenomenon, among them our theory that various tissues of these worms originate from genetically different eukaryote cell lines (or perhaps from different prokaryotes), and the observed genetic differences of these tissues reflect the ancient mechanisms of the origin of the multicellular eukaryotes. When we have more data concerning the gene sequences controlling ontogenesis, this hypothesis can certainly be investigated more closely.

Ontogenesis of multicellular organisms: an autogenetic process

Whether derived from prokaryotes or eukaryotes, all multicellular organisms undergo an ontogenetic process. It seems worthwhile to examine this process in the framework of the autogenetic model.

The overwhelming majority of multicellular organisms start their life as a single cell, and ontogenetic development of this cell produces the given organism's mature adult form. Ontogenesis is generally regarded as a process emphatically different from the evolution of the species. But the idea that the same laws of nature play a decisive role in both processes, ontogenesis and phylogenesis, merits consideration. That is, *ontogenesis is an autogenetic process.*

The zygote goes through a series of replications. Functional connections develop immediately between the "components" produced as a result of replication. The components are able to influence the probability of genesis—the further replication—of one another. Compartmentalization also can be observed. The activity of the cell population in the liver or brain of higher animals is substantially coordinated and is partly isolated within the organism. Finally, the formation of the complete organism can be regarded as a convergent process during which the organism is becoming a replicative unit, and, as such, for a long time it is capable of conserving its character in a more or less unchanged form. (This is true only with one reservation, because as we have discussed in the previous chapter only two individuals of the opposite sex taken together can be regarded as a replicative unity.)

The energetic changes observed in the course of ontogenesis support this notion (Zotin and Zotina 1967, Prigogine et al. 1972). In the early embryonic phases there is a detectably higher entropy production than in the later phase. The adult organism is in the state of a minimum stable entropy, which can only be changed by an accidental injury and a subsequent regeneration or by neoplasmic transformations, etc. Concerning the mechanism of ontogenesis, a very simple model can be conceived, based only on replication and on the mutual inhibition of the replicative function, thus regarding the development of the zygote as a true *autogenetic process* controlled by predestined constraints built into the genome. This assumption can be recognized as a reformulation of Haeckel's "biogenetic law." It seems that recent theories of ontogenesis based on irreversible genome repression are moving in this direction (Caplan and Ordhal 1978). It is very interesting that while in molecular evolution the self-organizing process came to a stop with the appearance of the cell, in multicellular organisms it occurs repeatedly during the ontogenesis of each generation.

As soon as the new compartment of the multicellular organizational level—the organism—has reached the phase of identical replication, the new compartment becomes a *replicative unit*. The information necessary for its replication is carried by its components: the cells. The effective range of constraints existing among the cells is on the order of meters. Compartmentalization of the autogenetic process on the multicellular level also has produced units of the same scale; that is, the volume of most organisms is below a few cubic meters. The organismic compartment also has *converged;* its control is almost perfect as a result of the mechanisms of identical replication. The fidelity of replication of its components is extremely high, and the functional cooperation of the components is highly differentiated and occurs in a hierarchical system.

Macroevolution

To define the exact nature of the components that emerged during the evolution of multicellular organisms is not a simple task. We can regard multicellular organisms themselves as components, although individually they are not able to replicate. Nevertheless, organisms form and decay in great number, and their mutual interactions start further evolution (the creation of newer organizational levels). Species also may be regarded as components because their inner structure and genotype are extremely stable (Mayr 1963). They can persist for millions of years under unchanged conditions. Changes in the environment naturally select among the species; some become extinct, while others change or produce new daughter species. Their average life span varies between 3 and 12 million years (Stanley 1975). Species originate and become extinct, so the definition of "components" can be applied. The process traditionally called macroevolution covers the complex changes involved in the origin and extinction of species (Simpson 1953). The early theories of evolution could not embody all problems, such as the origin of life, its present maintenance, macroevolution, etc., in a single unified model. Even Darwinian theory was concerned only with the origin of species. Theories of life's origin were constructed so that contradictions with the theory of macroevolution were carefully avoided, and, if possible, they used the same explanatory principles, such as competitive selection, that proved successful in treating the problems of macroevolution. This meant that much more emphasis was put on problems of macroevolution than on other problems. This is quite understandable as it was very important to provide an explanation for the origin of presently living species, if only for emotional reasons, and this has greatly influenced the direction of research. Our system-science approach is necessarily more general and concise and cannot follow the traditional path. Our aim is not to clarify the origin of a given species or even a whole phylum, but to attempt describing the system process in which such important but subordinated events necessarily occur. Therefore, we do not treat macroevolution in detail. We only continue our analysis of the formation of organizational levels with the autogenesis of ecosystems. Some important questions of evolutionary research also will be treated, among them the recent crisis of neo-Darwinian theory, in the light of our own model.

4 The Evolution of Ecosystems

With the development of multicellular organisms as a new structural unit, a new component capable of energy absorption and of establishing contacts with similar units has evolved. This new unit is the building block of organizational units higher than itself. Thus a new level of evolution, a new zero-system, has developed.

Ecosystem as a component system

There is no generally accepted ecological theory, and sometimes even the definition of ecological systems is debated. Usually the community, i.e., all living organisms found in a given area, is regarded as an ecosystem. Others include the various abiotic features of the environment. As discussed in the first chapter, the various system definitions may serve different goals, and the semantic arguments around the definition of systems are of no great importance.

If samples were taken from areas of various size on the Earth's surface, and if all living organisms in these samples were determined, it soon would be seen that there is a regularity in the kinds and number of organisms, partially depending on the size of the sampling area. Some of the regularities would be characteristic "patterns," reflecting the fact that individuals of certain species are more frequent in the vicinity of certain others, and there also would appear regularities concerning the physiology and behavior of organisms. There are, for example, certain stable ratios between predators and prey animals, or between the biomass of photosynthesizing plants and the consumers that use the energy and material of these primary producers through complex food networks. The science of ecology tries to find and analyze such rules. In any case, we may confidently state that the ecosystem or the community of a given area can be considered as a *component system*. We certainly may define individual organisms as components, regardless of their being simple soil bacteria or complex multicellular organisms, but it also is possible to consider populations or even whole species as components. Populations and species should be considered as higher organizations, anyway, although this is hotly debated by several authors, especially in the case of species. However, it seems very easy to find good evidence for the relative autonomy of the populations and species. A species, for example, can be considered as a community of "genealogical descendants." Its components have a common organizational past. On an evolutionary time scale species respond to environmental effects; the whole complex of components reacts by adaptation through the mechanisms of genetic recombination. (See Ghiselin 1981 for a detailed treatment of this question.) Here we only want to call attention to these problems, but we cannot go into details. For the application of our autogenetic model it is enough to consider the organism as a component.

Mutual interactions of organisms are described very well by food networks. The effect of the flow of matter and energy through ecosystems in regulating the replication of various organisms also is well known. Organization and complexity are considered especially important problems (Margalef 1963). If various communities are considered from the aspect of the diversity of species involved, or the material and energetic interactions among species, it becomes clear that a special concept of ecological complexity is needed. When the life of an area is destroyed either by natural catastrophes or experimentally, it gradually becomes colonized, and a new ecosystem builds up. While the return of the same species in the same numbers is quite rare, some general trends in the dynamics of repopulation can be found. In the new ecosystem there are usually fewer but more productive species, which produce an enormous amount of biomass with a less than optimal efficiency in using energy. In the long run new species join the food network, and the use of energy becomes increasingly effective. Most of the material flow makes a closure, conserving the elements. In the Amazon rain forest, for example, as much as 50 to 80 percent of the water is recycled (Salati and Vose 1984).

Simple and complex communities are well differentiated by the intuitive ecological concept of "maturation" (Margalef 1963). A new community is immature, but one that has lived for a long time, consists of a large number of species, and exploits many energy-saving and recyclic processes is regarded as "mature." The development of "maturation" is called succession; it is the temporal succession of populations living together, which reaches a terminal association through several intermediate stages. The final union is an equilibrium with climatic and geological conditions; it is stable and perpetual. The ecosystem evolving in the course of succession is characterized by the initiation of energy and material cycles (Duvigneaud 1967), which have an order-increasing effect just as do those at the molecular level of evolution (Morowitz 1968). In the course of succession the mutual control effects of populations are enhanced (MacArthur 1971). The degree of control is at its maximum in communities approaching climax, i.e., terminating succession. At climax the greatest part of absorbed energy is devoted to maintaining the system, which persists for a long time (Odum 1969).

Succession must be distinguished from evolution. First, it occurs much more quickly (a couple of hundred years), and it is not accompanied by evolutionary changes (origin of new species, for example). Succession can be considered rather as a process of regeneration, which compensates for the harmful effects of fluctuations in the environment by restoring the original composition of the ecosystem, the one that is most suitable for the given climatic conditions.

There are many studies on the organization of stable ecosystems. Why are a particular set of species living together and why are they living together in the way that they do? Many simple questions can be raised that are difficult to answer. Why are communities formed of a definite number of species? Why is it that new species do not settle? Many models have been worked out to answer these and similar questions. It has been shown that the effect of competition on the reproduction of populations of various species can be modeled by simple linear differential equations (Lotka 1925, Volterra 1931). Every species has several environmental requirements (temperature, humidity, food, etc.) for successful reproduction. In addition to these parameters the reproduction of most species is influenced by the presence of other species, among them those that are similar enough to compete for common resources. The functional system comprising all living conditions is expressed by the concept of *niche* (Hutchinson 1978). If competition is excluded, a multidimensional coordinate system can be created in which the axes represent particular living conditions. An abstract, multidimensional "space" can be defined in this coordinate system, which covers the essential conditions for a given species. This is called the *basic niche*. If the niche is drawn so as to include competition with other species, then a "realized" niche results. By using the niche concept, several of the above questions can be successfully answered. For example, it is well known that two species of identical needs cannot live together permanently. If somehow they are brought together, they start competing, and the species with the lower reproductive success will be excluded from the community. This phenomenon is known as the principle of *competitive exclusion*.

The niche concept is a good tool for explaining certain evolutionary phenomena. It has been observed that similar species, living in the same location and competing, can change in the very properties that are important in their competition, and the direction of change is opposite. This results in separation of the common niche and a decrease or a complete halt of competition.

Since the most important role in determining the living conditions of a given species (besides some basic abiotic conditions) is played by *other organisms*, a niche obviously is not only a functional description of a given species' needs but, just as importantly, a description of possibilities for life represented by a given ecosystem. This line of thought has led to the concept of the "empty" niche, which can be occupied by a species of a given set of needs. If such a species does not exist or somehow has been exterminated, a niche remains empty. In spite of this concept's abstract nature, it can be used in concrete cases. If appropriate data are given, the proportion of empty niches can be calculated; e.g., it has been estimated at 12 to 54 percent in modern ecosystems (Walker and Valentine 1984).

A certain dynamics also follows from the empty niche concept: a given ecosystem's functional interactions continuously create conditions that would make colonization by certain species possible. If a new species is incorporated, then not only the previously empty niche will be filled but the whole ecosystem's parameters will change. This also will change the niche structure of the ecosystem; i.e., new empty niches may be opened or old ones destroyed. This means that from an organizational point of view the ecosystem is not closed. Species not only can enter the ecosystem but can die out for various reasons. Many observers have shown that a given ecosystem can never get into an end-state concerning its species composition. First observations were made on small islands that were repopulated following natural catastrophes (MacArthur and Wilson 1963). It was found that small, isolated areas become repopulated very quickly; a succession process starts, but soon an equilibrium is reached between the number of colonizing and disappearing species. This is called *diversity equilibrium* and is explained by the assumption that species occupying a mature ecosystem form small populations, and the number of individuals in populations shows a binomial distribution. Thus it may occur that a given species' number of individuals will fall below the minimum necessary for survival, and the species will die out from the ecosystem. As the number of niches in a given community is finite, equilibrium between the number of incoming and disappearing species is quickly reached, and the system settles into a state of equilibrium.

It also was supposed that an equilibrium diversity is reached not only in succession but in evolutionary processes as well. In a mature ecosystem some species may become extinct forever, and new species may appear by *evolutionary mechanisms*. As an equilibrium number of species can be found on a whole continent, or in the whole biosphere, it can be assumed that origin and extinction of species are in dynamic equilibrium (Levinton 1979, Wright 1982). The task of various evolutionary theories is to explain how the processes leading to this dynamic equilibrium are connected with particular phenomena of macroevolution.

Autogenetic model of the evolution of ecosystems

When an ecosystem meets the criteria of a component system, the autogenetic model can be applied to it. Various unicellular and multicellular organisms can be considered as mutually interacting components of the system; thus functions emerge that create the higher structures of the organization's ecological level—namely, communities. Such a community (even the most primitive one) represents a new organizing agent; it is a system precursor and promotes replicative and competitive selection in the system. As a result, the system's components change in such a way that their functions become increasingly coupled (the process of coevolution). An increase in the replicative information concerning communities begins, together with convergence and compartmentalization. In the process of convergence among the community's populations, feedback relations appear (Pimentel 1961, Margalef 1972) and increasingly complex food networks evolve, which functionally correspond precisely to our supercycles. Moreover, ecological control processes analogous to compartmentalization have been discovered (Holling 1976).

The most developed product of autogenesis at the ecological level is the community. The range of forces contributing to the development of various functions is somewhere between a few centimeters and several hundred, occasionally several thousand, kilometers. The size of the ecological compartments also fall in this range. Mechanisms for identical replication have not yet developed. Therefore, fidelity of replication in various ecosystems is not very high. The low level of replicative fidelity allows for the continuous origin and extinction of species. More precisely, origin and extinction of species are basic processes, and yet no controlling system has been formed that would ensure a stable distribution of species in the ecosystem.

The notion of information also is used in ecology, and the quantity of information is calculated from the number of species and individuals (Odum 1969). The notion of parametric information introduced earlier can be used to express the distribution of species and individuals developing without the mutual influence of populations under the sole effect of climatic conditions. Naturally, in real life there is no structure (with the possible exception of monocultures) that contains 100 percent parametric information. The modification of components appearing as a result of the interaction of different populations can be expressed by the *functional* information content. Associations at the beginning of succession have a high parametric and a low functional information content. The latter gradually increases in the

Figure 15 Beak types of the Galápagos finches (after Lack 1967).

course of succession, reaching a maximum at the climax when it is mostly *replicative* information.

It can be assumed that there is a connection between the niche concept and replicative information. For the replication of a given organism, aside from the information of its own genome, all necessary conditions are described by the concept of niche. The community carries this information without the actual presence of the given organism; therefore, this information must be *a part* of its replicative information. The functional (replicative) information embodied in the niche is an active factor in evolution, as shown by several examples of radiation and the emergence of morphological analogies during evolution.

The evolution of the Darwin finches of the Galápagos Islands shows how species, if they occupy an empty niche, change in their morphology and behavior to fit precisely into the community already formed. The Galápagos Islands are far from mainland South America, and many species of finches developed there from a common ancestor, adapting to various empty niches mostly by changes in their beaks (figure 15). Some finches evolved beaks suitable for feeding on very small seeds, while others evolved to feed on larger seeds. Parrot-like beaks developed in species adapted to feed on fruits and buds. Some species have beaks like forceps for feeding on small insects. The most interesting species colonizes an empty niche usually occupied by woodpeckers on the continent. This species has not evolved a new kind of beak but has adapted behaviorally by using thorns from cactuses to pick out worms under the bark of trees.

The real agent enforcing changes is the *information* embodied in the members (components) of the communities (compartments) i.e., the same functional effects that established the niche. The organism's replicative information and the niche's information content can be viewed as cast and mold; they determine and complete each other.

Evolution of the biosphere

With our discussion of the autogenesis of ecosystems we have reached the highest level of evolution on Earth. Inaccurately replicating ecosystems themselves also are subject to selection (Dunbar 1972). The final level of evolution on Earth is formed by the participation of ecosystems as components. The entity at that level is the biosphere itself. It seems as if evolution has come to a halt mostly because—in contrast to the lower levels—the component at this level is *a single entity*.

Organizational levels formed like an onion from the molecular level to the biosphere level. The most elementary building blocks, the atoms, form molecular components, and their autogenesis forms the cells. Organisms are formed from cells as components, and their autogenesis has led to ecosystems. The interaction of ecosystems manifests itself in the biosphere. In this account, organizational levels having very limited autonomy, such as cell organelles or the organs of the higher organisms, have been omitted. We may consider these as *abortive levels* of organization. Organization proceeded from "below" during the biosphere's evolution, but it is quite possible that even during the cells' autogenesis functional effects acted on the level of the whole system and can be considered as agents of the highest organization. Such an effect was possibly expressed in the composition of the atmosphere discussed earlier. This means that although levels of evolution can be logically separated, in reality the autogenesis of the various levels started at the same time and strongly influenced events at other levels. The future of the biosphere can only be considered if we have already discussed the other evolutionary processes accompanying organismic evolution. With the emergence of multicellular organisms, not only has ecological evolution started, but evolutionary sublevels such as neural, cultural, and technical evolution have formed. After discussing these we will return to the biosphere.

5 Some Unsolved Problems of the Evolutionary Theory and the Autogenetic Model

Neo-Darwinism

When Darwin was writing his famous book, the science of genetics, which provided most of the practical proofs of his theory, had not yet been born. In spite of this, Darwin was able to articulate his ingenious theory. Regarding the basic mechanisms, Darwin considered individual differences—that is, variability and environmental selection—the two most important factors of evolution. He assumed the gradual accumulation of minor useful changes—that is, in recent terms he was a "gradualist."

After the discovery of Mendel's law at the turn of the twentieth century, a great debate started between gradualist "biometricians" (K. Pearson and W. F. R. Weldon) and "Mendelists" (whose leading representative was W. Bateson). Biometricians worked on the inheritance of traits that show continuous variation, such as height, weight, etc., and they could not find the trace of discrete genes in the phenotypic distribution of such traits. Mendelian genetics also was supported by the discovery of mutations (de Vries 1901), for the effect of various mutant genes was great and their inheritance followed Mendel's rules.

Mutation was an example of a sudden change of traits. Studying continuously varying traits provided useful data for the explanation of evolution through the accumulation of small steps. A sudden change in traits was emphasized by the Mendelists; therefore, they were considered "anti-evolutionists" for a long time.

The exact definition of Mendel's law was the starting point for further discoveries in genetics. The most important among these was the Castle-Hardy-Weinberg law (Castle 1903, Hardy 1908, Weinberg 1908), which is the basic tenet of population genetics. This law states that in a large "panmictic" (randomly breeding) population without outside or internal disturbing factors the distribution of allelic variants is stable.

The great founders of population genetics analyzed the factors that might disturb this stable distribution. Haldane (1924), in his famous first publication, examined the effects of selection in the cases of lethal, linked, and semidominant genes, considering inbreeding and the effect of overlapping generations, etc. Wright (1931) studied the role of random fluctuation in gene frequencies using mathematical models, and he found that random fluctuations may become amplified by the effects of local selective factors, and thus they can change the equilibrium of gene frequencies. Later, this mechanism was termed "genetic drift." Finally, Fisher (1930) showed that Mendelian rules are consistent not only with the Darwinian theory but are the most important proof for it. The expression of continuous traits can be explained by the effects of many so-called minor genes.

The works of Haldane, Wright, and Fisher started *neo-Darwinism*, which is a combination of biometrics, Mendelian genetics, and the Darwinian theory of evolution. Its pinnacle finding is that evolution is simply the slow but continuous change of the gene frequencies of populations.

The success of neo-Darwinism was not complete, mostly because "naturalists"—biologists who were interested in analyzing evolution by studying living populations and fossils—withdrew behind the fortifications of their own methodology. Naturalists mostly have studied the diversity of species and the mechanisms of development of this diversity (Mayr 1982). They recognized the role of geographical isolation in the origin of species, and thus the isolation of a small population for a longer time was thought to be a precondition for the origin of a new species. It was assumed that mechanisms that secure the identity of the new species after the original isolation ends are formed during isolation. These two branches of evolutionary research united in the 1930s within the framework of the "*synthetic theory*." This framework was created mostly by the works of Huxley, Dobzhansky, May, Simpson, Stebbins, and Rensch, who were able to recognize that these two branches of research could be united into a new paradigm. The basic tenet of the synthetic theory is that the changes of allele frequencies in microevolution and processes of macroevolution are connected, and that all evolutionary changes can be explained on the basis of the same principle. Evolutionary changes at higher organizational levels can be explained by microevolutionary processes.

Neutral evolution

The synthetic theory was challenged by the development of molecular biology. While the naturalists were using overt morphological traits in describing species, the advent of molecular biology made possible the use of fine structures in giant molecules in evolutionary research. It has become possible to study the homogeneity of a population established by morphological traits using the fine structure of cell proteins of the given species. Enormous variations have been discovered in seemingly homogeneous populations. Within species, numerous variants of enzymes and other proteins have been found. These variants were distinguished by changes in amino acid sequences, and correspondingly there are many allele variants of the respective genes within the population. This type of variability is called *molecular polymorphism*.

It soon was found that the catalytic activity of an enzyme was rarely influenced by polymorphism; that is, variants were *functionally equivalent*. This finding caused serious problems for the neo-Darwinian theory. From previous definitions of the theory it followed that all traits of an organism were the result of natural selection; therefore, polymorphism of proteins also should be produced by it. The only problem was that attempts to find conditions responsible for the evolution of polymorphism were unsuccessful. For a long time it was believed that this was only a matter of finding *adequate* data, and the discovery of special environmental effects responsible for polymorphism was very much hoped for.

But soon it became clear this was not a methodical problem but basically a theoretical one. Following the discovery of the genetic code, it was found that certain "silent mutations" exist, which are changes in nucleotide sequences in the DNA that do not change the corresponding amino acid sequence in the respective protein because of the redundance of the genetic code. This was an example of genetic variability without the slightest consequence to the properties of the organisms. From these phenomena a unified theory, named the "neutral mutation theory," was born (Kimura 1969, 1976, 1977, King and Jukes 1969, Kimura and Ohta 1971).

According to the neutral theory, not all parts of a molecular structure are equally important for its function. There are parts of the protein sequence that may be occupied by several different amino acids, and their change is "neutral" from the point of selection; that is, the fate of these changes is not influenced by natural selection. It also was found that some parts of amino acid sequences cannot be changed, or else they can be changed only rarely, and these play a very important role in the specific function of the protein molecule. For example, Kimura showed that more changes may occur on the hemoglobin molecule's surface than in the location of its inside "pocket," which plays a role in binding oxygen.

Protein polymorphism also can be studied among species, and it has become a fine tool for the investigation of evolutionary distance between different species. The earlier that species have been separated from a common ancestor, the larger the difference between their protein structures.

From these results it follows that species do not have proteins and nucleic acids of stable and final structures, but at the top of the molecular hierarchy DNA is continuously changing, and, correspondingly, RNA and proteins also change. Some of the changes are unfavorable and are eliminated by selection, while neutral changes survive and accumulate, changing the molecular composition of a given species. The rate of change is relatively fast. According to calculations by Kimura, the rate of change in mammals is 2.5 nucleotides per year. This means that each nucleotide of the DNA of living organisms has changed at least four times since the origin of life (Kimura 1977). This, of course, is an average value. The functionally important nucleotides are changed at a slower rate.

Discovery of neutral mutations has changed the earlier static image of the DNA. On the evolutionary time scale the DNA's structure is a dynamic, one-dimensional "ocean" of randomly changing nucleotides in which there are stable "islands" of the nucleotides responsible for the proteins' functional sites. This image can be extended to the macroscopic properties of organisms. It has become questionable whether all traits result from natural selection, as the existence of neutral traits with no adaptive value is possible and requires explanation.

Usually *genome* and *gene duplication*, which play a very important role in evolution, are not included in the theory of neutral evolution (Ohno 1970). However, increases in the genome size and development of new functions probably are due to genome duplication. In gene duplication the genome is extended by a piece of DNA, a copy of an existing gene, which is neutral regarding selection. Its existence is neither advantageous nor harmful until the new piece of DNA gains a new function through subsequent mutations by which it becomes useful.

It is quite clear that neither the existence of neutral mutations nor evolution by gene duplication has followed from the synthetic theory; moreover, taking it rigorously, they are denied by it.

Punctuated evolution and stasis

The problem of neutral evolution was first raised by population geneticists, and in the 1970s it was raised by those naturalists (or, more precisely, paleontologists) who attacked the neo-Darwinian theory.

Eldredge (1971), Gould (Eldredge and Gould 1972, Gould and Eldredge 1977), and Stanley (1975, 1979) presented arguments based on paleontological evidence. They stated that traces of gradual change assumed by the neo-Darwinian theory cannot be found in the case of most species. Evolution as shown by fossils is characterized by sudden, great changes, and following *stasis* it is characterized by an equilibrium state lasting for millions of years, during which species do not change or change only slightly. The most important mechanism of macroevolution is the selection among *species*. This theory, which was termed "punctuated equilibrium," has created an enormous confusion in evolutionary research, mainly because in support of the theory paleontologists presumed genetic mechanisms that for a long time have been doubted by geneticists. Such a mechanism, for example, is Goldschmidt's (1940) theory of macromutations resulting in "hopeful monsters." Goldschmidt assumed that in special cases large-scale genetic changes may be induced individually that can produce gross morphological changes (new organs, etc.) in one generation. However, it is well known that large-scale changes usually induce nonviable monsters, and a "hopeful monster" able to evolve any further has never been found.

The concept of sudden changes also has been criticized widely because 50,000 to 100,000 years is rather long for a geneticist to assume gradual changes, while this same period is very short on a paleontologist's time scale. Stebbins (1982) argued that gradual changes at the microevolutionary level permit large, sudden morphological changes at the macroevolutionary level. Others argue against the concept of stasis (Schopf 1982) by pointing out that only fossilized traits serve the paleontological models, and physiological and behavioral changes that usually accompany the origin of species are disregarded by paleontologists. There is a dangerous possibility that the constancy of a species is restricted to the "fossil species," and the constancy of the real "biological species" could have been much shorter in time. According to Charlesworth et al. (1982), punctuationists do not present any new facts requiring revision of the classical theory, and their findings are entirely consistent with the neo-Darwinian theory. The possibility of sudden changes was mentioned by one of the synthetic theory's founders and was termed "quantum evolution" (Simpson 1944). Wright (1982), one of the founding fathers of neo-Darwinism, thinks that the theory, especially his variant, the "shifting-balance" theory, can account for the paleontological findings. According to the shifting balance hypothesis, large populations quickly reach an equilibrium state during evolution and are maintained practically unchanged for a long time by stabilizing selection. However, small, isolated populations can shift to another equilibrium state through the mechanism of genetic drift, and so new species may originate. It is emphasized by Wright that the most important factor in the creation of a new species is the structure of the ecological niche that so far has been largely ignored by evolutionary theory.

Finally, quite ironically, a paper has been published that attempts to prove that even Darwin was a punctuationist (Rhodes 1983).

In a critical assessment of the arguments a firm conviction seems to develop, namely, that on a geological time scale the origin of a new species is a rapid process, the phenomenon of stasis is real, and the formulation of a new evolutionary theory is required that considers different levels of organization as the living world's basic structure and can provide a unifying explanation for processes of both micro- and macroevolution.

Adaptation, fitness, and selection

Adaptation is one of the synthetic theory's most important and, recently, most debated concepts. In loose terms adaptation is an optimization process in which a population's genetic architecture changes under the effects of environment in such a way that the changes favor survival and reproduction of the individuals within the population. In other words, in possession of the parameters of the environment, the optimal properties of an organism can be calculated, and their correspondence to actual traits can be analyzed (Lewontin 1979). Neo-Darwinian theoreticians have tried to turn this loose definition into an exact mathematical model, operating with gene frequencies and the Darwinian concept of "fitness"—the phenotype or genotype's probability of survival and rate of reproduction. Fitness is a consequence of the relationship between the organism's phenotype and the environment in which the organism lives; so the same genotype will have different levels of fitness in different environments. It follows from the theory that during adaptation the average fitness of alleles increases. The adaptation concept's critics argue that within this framework adaptation is taken for granted, and investigations are concerned with the ways in which various organisms adapt rather than with first establishing the fact of adaptation (Lewontin 1979). It is beyond doubt, and experimentally demonstrated, that certain traits are formed by strong selective forces. But this does not prove that *all* traits have originated as a result of selection and that only selection may have had a decisive role. Most genes affect many traits, i.e., they are "pleiotropic." It is difficult to assume that a gene influencing many different traits can change during adaptation in a way that is optimal in all respects. It is more plausible to assume that traits also can be formed by random effects or can be the by-products of selection for other adaptive traits.

The question also arises: what is a trait? If a fruitfly is examined, tens of thousands of traits can be described because the sole criterium of a trait is that it should be distinguishable. However, the various traits obviously are not entirely independent. Chromosomes of the fruitfly carry a definite number of genes (something over five thousand) that are responsible for the development of all traits. The "real" number of phenotypes is somewhere between four thousand and several tens of thousand, but we do not have an appropriate systematics for the phenotypes that would provide an adequate estimate.

Many reasonable criticisms have fallen on the adaptation concept, but one cannot give up the conviction that the environment *influences* the organisms and that the latter can occasionally adapt. Nevertheless, the fact of adaptation is always questionable and requires solid evidence.

Many authors also have criticized the definition of fitness. Beyond the fitness concept of population genetics, evolutionary theories use a concept referring to the whole organism, meaning the fitness of the individual based on its reproductive success. However, it was shown that this definition of fitness is tautological and overlooks empirical content (Kimborough 1980, Michod 1981) at least in the form that defines the organisms of highest fitness as those that have survived during evolution. Recently, some authors have tried to avoid the tautological definition by giving an empirical content and using an ecological interpretation of adaptation (Kimborough 1980). Ecological conditions are sought that influence the life and reproduction of organisms and to which they must adapt.

Debates concerning fitness also raise the problem of the unit of selection. The concept of inclusive fitness was developed by sociobiologists (Hamilton 1964). It includes individual fitness (expressed in the number of offspring) and fitness represented by the reproductive success of relatives carrying similar copies of given genes. Inclusive fitness is rather difficult to deal with at the level of individuals. Therefore, Dawkins (1976) promoted the idea of *genes* as the final units of selection.

In his excellent and daring analysis Dawkins introduced the concept of *replicators*, which he regards as the unity of selection. Replicators are entities that interact with their environment and originate by *copying*. To Dawkins, genes are replicators, but organisms are not. Selective processes influence the structures of replicators. We note here that his concept resembles our concept of replicative information, introduced about the same time (see Csányi 1978). Although the replicator concept solves several problems concerning the levels of organization, it does not answer the questions raised in connection with adaptation and fitness.

The most profound criticism of the traditional concept of adaptation and fitness was given by Ho and Saunders (1979), who showed that it is rather unjustified to suppose the necessity of highly selective conditions for the origin of new species. They find just the opposite; i.e., *relaxation* of selection favors an increase of variability and thus the origin of new species. Their idea of the complex nature of selection will support some of our arguments later in this book. They separate stabilizing selection preserving various existing forms from other selective mechanisms that serve to secure the consistency of structures with life. According to Ho and Saunders, the most important source of evolutionary change is interaction between environmental conditions and the epigenetic processes of ontogenesis.

There are some remarkable attempts to use recent discoveries in microevolutionary processes to explain macroevolutionary phenomena. The finding of neutral mutants and observations on the continuously changing nature of the genome reveal phenomena that must somehow also be expressed at the macroevolutionary level. One popular explanation is the "Red Queen" hypothesis of Van Valen (1973). This theory states that an advantage gained by some species through adaptation is expressed as a degradation of environmental conditions for the ecosystem's other species. The degradation of environmental conditions, in turn, promotes adaptive changes in these species that produce further changes in those that are influenced. In such a way a continuous change, a kind of "running-in-place," is created, which is reflected in continuous changes of nucleotides in the DNA at the microevolutionary level. This hypothesis agrees with ecological models predicting the continuous origin and extinction of species (MacArthur and Wilson 1963).

The meaning of adaptation, fitness, and selection in the autogenetic model

The problems discussed above point to the most fundamental insufficiencies of current evolutionary thinking, namely, that it separates evolutionary processes from each other and ignores the levels of organization and the fact that *the whole biosphere is a systemic unity,* the changes of which can be understood best in the systems science's theoretical framework. What we are attempting here is to provide just such an interpretation of these concepts.

Theories of classical biology always attempted to find explanations for biological phenomena starting at lower levels of organization and then moving upward (if they considered organizational levels at all). Such an analysis was suitable for only *causal* explanations and unsuitable for an ultimate understanding. To distinguish between two kinds of explanation, we take the example of enzyme reactions. An enzyme molecule can be regarded as a specific constraint, influencing the dynamics at a lower level (chemical reactions). Analysis of the enzyme structure and mechanisms of enzyme reactions provided explanations of *how* enzymes function—that is, the mechanism of enzyme catalysis—but this analysis is unsuitable *in principle* for clarification of *why* enzymes have been created, why exactly certain kinds have been created, etc. These questions can be answered only at the organizational level of the cell. In the same way at the organismic level the dynamics of the organisms—their mutual interactions—are constrained by the ecological level, and these constraints can be understood only in the framework of ecology. Interpretation at the organismic level, or the even lower molecular level, is completely impossible.

As has been stated by Ho and Saunders (1979), the levels of organization show a *temporal hierarchy*. Organizational levels can be separated by the temporal characteristics of the dynamics of their components. Changes affecting the whole biosphere (at least in the course of known evolutionary history) have occurred over 1 to 10 million years. Changes shaping the ecosystems act on a scale of a hundred to a thousand years. Processes in organisms take seconds; the time scale within the cell is 1/1,000 to 1/10,000 second. These ranges can be distinguished very well, and they indicate that controlling effects act downward and not vice versa. The energy consumption of a multicellular organism is determined by its organismic needs (defense or work for food, for example) and not by internal microprocesses (between given limits, of course).

One characteristic of the whole biological system has caused many problems in evolutionary thinking and has been misunderstood many times. Namely, there are many organizational levels of the biosphere, but *every change is fixed at the lowest molecular level in the* DNA because of the very nature of the component-producing processes. Constraints at the various levels influence the dynamics of lower levels, but the description of those constraints is embodied in the components of the lowest level. (Here we mention only that neural and cultural levels of evolution are different in this respect, as will be treated in detail.) It is quite clear that the behavior of the system would be very different if, for example, the constraints at each level were entirely independent of constraints at other levels. There are many consequences of this characteristic, and some will be discussed below.

Occasionally, constraints at the various levels can change. Some of these changes are only temporary, as the constraints themselves are reproduced by the component-producing processes. Thus permanent changes can occur only if the component-producing processes change in a way that is fixed in the *genome*. There is not much sense in speaking about selection of individuals or species, because no selection can change the system permanently unless the genome is changed. Therefore, the real unit of selection must be the gene, as Dawkins has argued.

Adaptation can be considered the process of fixation within the genome of changes in constraints at the higher levels. As a consequence, the only processes that are adaptations are those that contribute to the formation of the specific constraints. Chemical reactions in the cells are controlled by constraints, but not all chemical processes (not all chemical characters) are the consequence of constraints. The same is true for organisms. Interaction of cells creates the organism, and this process is controlled by the specific constraints of a higher level of organization. But not all traits are consequences of these constraints. Various structural solutions, forms, etc., have occurred as a result of the cell's spontaneous dynamics. Constraints appear only where they are needed for the higher level of organization; in other words, adaptation cannot be taken for granted.

From our interpretation of adaptation it follows that fitness may have only a relative interpretation. Since there is a lag in all adaptation processes, fitness of a population can be calculated as the ratio of genotype frequencies and the mean frequency of the best possible genotypes in the given environment. This fitness value would show the degree of deviation of a given population from an ideal state.

Our concept of fitness, perhaps more exactly than the present one, expresses the population's actual state on its course of adaptation to the present environment. (We put aside the question of whether the genotype of maximum fitness in a given environment can be determined.)

Before interpreting selection we must deal with the concept of species. Species can be regarded as components in the autogenetic model, but the level of species cannot be regarded as a separate level of organization. Species, just as organs of the multicellular organisms, have only a low degree of autonomy. For this reason we regard species as *compartments*, with individuals as their components, that mutually interact through reproduction and genetic recombination, resulting in a more or less coordinated replication. Therefore, the species is a replicative unity. Since we consider the species a compartment, the origin of species is a process of *compartmentalization*. All effects that inhibit the free flow of genes among the populations of a species might lead to the formation of new subcompartments—that is, new species. Longtime geographical isolation, changes in the chromosomal structure, mutations inhibiting recombination, genome duplication, etc., all are these kinds of effects, possibly leading to speciation. The species compartment probably is created as a result of the rapid spread of new favorable genetical structures by the mechanisms of recombination and replication, and so a great number of similar components (individuals) are produced. At the same time division of the species into subspecies and then the origin of new species are epiphenomena because they result from obstruction of the free flow of genes.

Creative and negative selection

Most problems concerning natural selection are connected with the fact that natural selection is not a homogeneous process. In discussing molecular evolution we have argued that there are two main types of natural selection (competitive and replicative) that are very different with respect to their outcomes. Other researchers also have recognized that natural selection is not homogeneous. Ho and Saunders (1979) contrasted stabilizing selection, which is negative in its effects, with selection for viability. This distinction, even if it has not been elaborated in detail, is very similar to our distinction between competitive and replicative selection. Also worth mentioning is the idea of "autoselection" by Riedl (1978), which concerns the effects of already existing biological structures on their own selection. The stabilizing effect of competitive selection is well known; it has the primarily *negative effect* of eliminating new variants. (However, we cannot go into its details here.) Replicative selection, on the other hand, is a new concept, so we shall discuss its consequences in more detail.

Replicative selection's only negative effect is its elimination of changes that risk or inhibit replication—that is, division of cells or reproduction of higher organisms. Compared to this negative effect, its *creativity* has a much greater importance. Replicative selection permits every change that does not inhibit replication. The degree of freedom of replicative selection therefore is very high. There is an enormous variety and richness of forms, and all these are capable of self-replication. The creative nature of replicative selection can be shown satisfactorily. Its evolutionary creativity is evident to biologists, and sources for this creativity have long been sought. Since it has been supposed that natural selection is homogeneous and competitive, demonstration of competitive selection's creative nature has been attempted without much success. The variety of animals and plants specially bred by man are frequently mentioned as proofs of the creative nature of selection. However, these are not very good examples because the creativity of artificial selection, while it is a process still under outside control, is organized by man. Natural selection is often regarded as the effect of the environment, so it may be important to interpret the concept of environment within the autogenetic model. Environment in the autogenetic model is very simple. It consists of energy, the source and sink that make the autogenetic process possible. When evolution of a particular organism is analyzed, biologists tend to regard all other components of the biological system as the environment. This treatment simplifies the study, but it has the disadvantage of treating the environment as an *outside* factor. With this dichotomy of the organism and its environment, the most important characteristics—the mutual interactions of the biological systems—are excluded from our thinking. According to the definition of autogenetic process, the components of a system constitute a functional network, which continually renews the same network with an increasing precision; i.e., it self-replicates. Instead of the biological concept of environment, we are concerned with the functional network of components, and in that way the organism-environment division is replaced by a component-system dichotomy. There are far-reaching consequences of this approach. It is evident, for example, that the most important selective force—either replicative or competitive selection is involved—is the system itself (or, in terms of the autogenetic model, the system's replicative information). Components of the biological system are materializations of the constraints of organizational levels. The functional system of constraints is the system's organization. This organization is the active selective force behind both replicative and competitive selection. In that way the already existing organization is related to itself through positive feedback. The *self-organizing properties* of the autogenetic system are a direct consequence of this positive feedback.

Creative and negative selection at various organizational levels

Going downward, five major organizational levels can be distinguished in the biosphere.

Fifth level: the biosphere

This is the highest level. Therefore, there are no constraints that could influence the dynamics of this level's single component—the biosphere of Earth. Conditions of selection are determined only by the physical conditions of the system's existence: radiation from the Sun, chemical-geological conditions of the Earth's surface, abiological atmospheric processes, temperature, etc. Competitive selection also is absent from the biosphere's organizational level as there is only one component. Replication of the building blocks of the biosphere provides the nonidentical replication for the whole system. The biosphere system is open; its operation, provided it can replicate, has no constraints besides the physical conditions.

Fourth level: the ecosystems

Here, and on all lower levels, conditions and constraints will be distinguished. Constraints can be interpreted as imposed from the next higher level of organization. Conditions are either outside physical parameters or the effects of organizational levels above the next higher level.

There is an organizational level above that of ecosystems, and constraints on ecosystems arise from this level. These constraints are created by the totality of the biosphere, for example, the biologically produced free oxygen, carbon dioxide, water, etc. Constraints created by the bio-geochemical cycles, or by the space requirement of the various communities, also belong to this higher level. Conditions are created by local chemical and physical paramaters, leaving out of consideration which of those originate from internal interactions within the communities. The role of negative stabilizing selection is to promote adaptation of the system's components—the communities—to the constraints. Replicative selection is open and creative. Replication is in the nonidentical phase.

Third level: the organisms

The components at this level are the species, and individuals are the subcomponents within them. The specific constraints at this level correspond to the replicative information expressed in the niche networks of the ecosystems. Stabilizing selection acts according to the replicative information of the occupied niches. Replicative selection also is constrained and is subject to the replicative information content of the empty niches; therefore, its creativity is minimal. The conditions of the organisms' existence are determined by the chemical and physical parameters of the biosphere. Replication of the components is nearly identical.

Second level: the cells

The living conditions of the existence of cells are determined by the internal environment of communities. The specific constraints at this level are determined by the replicative information of the organisms.

Stabilizing selection is weak, but still detectable, and the replicative selection mainly concerns viability. Its creativity is extremely low. There is very little chance for a cell to change adaptively because all new variants must fit into the organism's existing organization. In the cases of free-living cellular organisms, this constraint is absent. Replication of cell components is almost identical.

First level: the molecules

Conditions are given by the organism's internal environment (or the physical environment of the community in the case of unicellular organisms), while constraints are determined by the replicative information located in the cell. Stabilizing selection can be detected among the molecules. Creativity of the replicative selection is almost negligible, although it is not zero, because creativity of the biosphere is manifested in the changes of molecular structures. The degrees of freedom of changes, however, are only a fraction of those at the higher levels. Structural changes are permitted only if they fit into the system of constraints and conditions created by all the higher levels. Replication of the molecular components is almost identical.

The organizational structure of the biosphere is *open from above*. The degrees of freedom increase upward, and at the highest level, aside from the external conditions, only the replicative selection of the components themselves restricts changes in the system.

Knowing the organizational levels, the *direction of evolution* can be determined. The evolutionary process starting from the first zero-system proceeds in the direction of maximizing replicative information contained by the system. Maximization of replicative information is realized through the convergence of the system's compartments on each organizational level. The physical space of the autogenetic compartments of subsequent organizational levels increases. The range of forces forming the compartments also increases on each subsequent level, as has been discussed (figure 16).

Evolution of the biosphere has not proceeded at a steady rate; sometimes existing ecosystems change abruptly. The dinosaur ecosystem, for example, had been stable for 100 million years when it was abruptly changed and followed by an entirely new ecosystem. A complex mammalian megafauna was formed that was no differ-

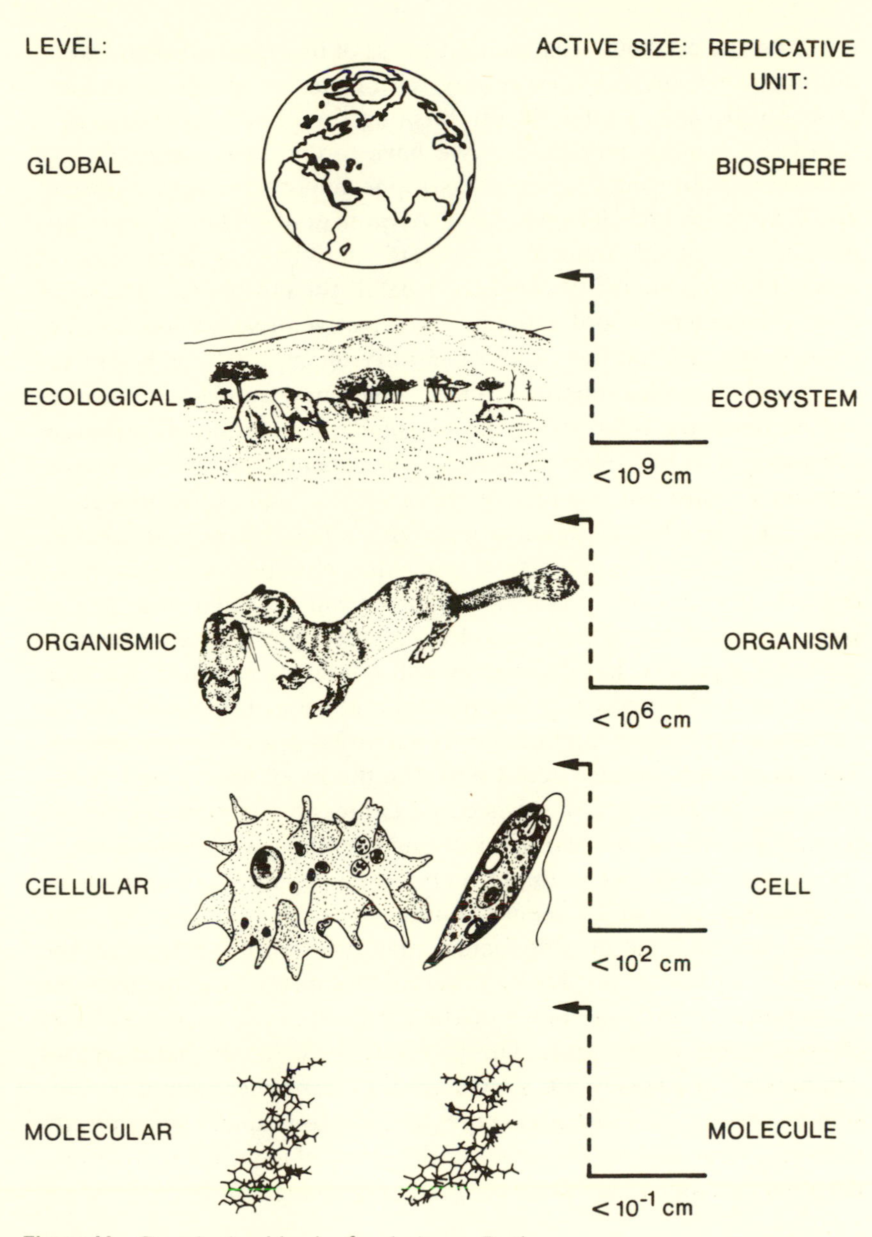

Figure 16 Organizational levels of evolution on Earth.

ent from the previous ecosystem in terms of its organizational levels, although its complexity, especially with regard to man's emergence, probably became greater. In establishing the direction and dynamics of the evolutionary process we have to consider *outside* factors influencing autogenesis (see catastrophe hypotheses by Alvarez et al. 1980, Smith and Hertogen 1980, Alvarez et al. 1984). If powerful, destructive, outside influences (volcanic activity, effects from outer space, drastic climatic change, etc.) befall the biosphere, then some of its components coordinated by a replicative organization may become destroyed and the system's *identity* changed. If conditions for life remain between tolerable limits, then a new autogenetic process begins, with the remnant of the biosphere functioning as a system precursor. The functions of existing components "search" for opportunities to close the replicative chains. New cyclic processes start, and a new replicative organization, with a new identity, is created. Obviously, the more complex the original biosphere was, the more complex the remaining system precursor can be, even in a case of large-scale damage. Thus the following autogenesis may result in a greater complexity, but it probably will not create a higher organizational level. Quite possibly the great complexity of the biosphere on Earth was created by random but frequently repeated outside effects. This mechanism would accord with the theory of punctuated evolution. The phenomenon of stasis could express replication at a higher level of fidelity by the ecosystems. Sudden changes accompanied by the origin of new species might reflect the temporary collapse of the replicative system and subsequent autogenesis.

Both internal and outside factors may have a role in forming the biosphere's present complexity. Internal factors promote the increase of complexity until the range of the controlling constraints reaches the system's physical limits. This probably is the factor that increases complexity in a steady but slower way. *Catastrophes* caused by outside factors probably increase complexity more rapidly—but unpredictably. If the system ever reaches the phase of identical replication, then only outside factors will promote any changes in it.

6 Neural Evolution

The essence of the forthcoming hypothesis is the assumption that animal memory is based on the interaction of *replicative components* formed by neural interconnections. From this assumption follows the development of a new type of evolutionary system. While the structure and features of DNA, the basis of biological evolution, are well known, there is only indirect evidence for the existence of replicating elements of memory. (However, it should be kept in mind that Darwin formulated his evolutionary theory without any basic knowledge of genetics.)

Energy flows in the multicellular organisms

The highly differentiated cells of multicellular organisms intensively exchange matter and energy. Almost 20 percent of the body's total energy requirement is used by the brain (Lajtha 1974). Only a small percentage of this substantial energy is necessary for the brain's electric activity, while most of it is used for the continuous chemical renewal of neurons.

The most intensive protein synthesis takes place in the brain; however, its exact biological function is not yet understood. The most essential feature of the matter and energy exchange determining the connections between groups of neural cells and between these and other cells is that its control is partly subject to *external factors*. These factors, the stimuli, open certain specific channels for the energy flow going through the nervous system. The more complex the nervous system, the more complicated is the route of energy flow, and the longer that excitation produced by external stimuli is stored in the brain's structures. In the autogenetic systems discussed so far it was evident that the energy flowing through a component system acts as an organizing force; in response, the system deviates from equilibrium, and its structure becomes more and more complex.

The effect of the energy flow through the nervous system is expressed in *animal behavior*. One of the most important aspects of animal behavior is the response to external stimuli. According to Lorenz, the most primitive kind of response is *instantaneous* (Lorenz 1973). Receptors of unicellular organisms essentially work in this instantaneous manner (Gould 1974), for the primitive organism and its environment can influence each other only simultaneously. In multicellular invertebrates a new type of response, *sensitization*, appears. In the course of sensitization the excited state of receptor cells, transmitted to the nervous system, evokes not only an immediate reaction within the nervous system but produces a lasting state of excitement. The organism reacts to the reappearance of the stimulus more immediately, more directly, and more effectively (Wyers 1976). The mechanism of *habituation* is essentially similar. Experiments on the sea snail, *Aplysia californica*, show that habituation to touching is based on the exhaustion of interneurons connected to the receptor cells. No similar exhaustion takes place even after tens of thousands of stimulations in the receptor cells. The ability to habituate involves the "disconnection" of repeatedly occurring stimuli, that is, the attitude of reservation and inactivity. These two complementary processes, sensitization and habituation, provide the basis for higher cerebral function. This triad—the complex of receptor cell, interneurons capable of sensitization and habituation, and the motor neuron—provides the simplest neural model of behavior. Study of the nervous system and the behavior of invertebrates points to the dominant role in behavior of neural networks generating autonomous reflexes (Wilson 1970) and inherited locomotive patterns (Bullock and Horridge 1965, Kandel 1976). The structure of neural networks is very stable and shows only insignificant individual variability (Hoyle and Burrolus 1973). The permanent changes induced by environmental effects such as learning and memory already can be detected in a rudimentary form, but with some special exceptions they play a secondary role. In other words, the organizational effect on simple neurons of the energy flow through the system is still negligible.

On the other hand, the neural activity of vertebrates is characterized by a continuous and irreversible "reorganization" of the neural network, which is induced by the environment and lasts from birth to death. Here, too, the receptor and motor structures are fundamentally invariant and genetically determined; neural structures capable of storing the energy modulated by information coming from the environment have evolved, reaching extreme complexity (Rakic 1975).

Models of the external world in the brain

For the simplest neural systems the external world consists entirely of a few stimuli. This external structure of stimuli becomes, so to say, projected into the nervous system through the processes of sensitization and habituation. The motor apparatuses are not activated directly and exclusively by external stimuli reaching the receptor, but by the integrated effect of the external stimuli reaching the receptors and an interposed "model" composed of interneuron-type networks. Together, the two shape and elicit the response reaction (Craik 1943). Mechanisms that adjust the internal parameters of the organisms to random or temporary changes in the environment can be found at every level of the living organization. These controlling processes constitute *physiological adaptation*. The nervous system is a special organ of animals' adaptation, created in the evolutionary process to coordinate the various processes of adaptation. It was brilliantly shown by MacKay (1951–52) that the essence of this adaptation is *goal-directed behavior*, and neither vitalistic nor anthropomorphic concepts are required to explain it. Goal-directed behavior has a simple cybernetic explanation if we consider the organism a system, with a number of feedback loops, capable of changing its own parameters.

The definition of goal "X" pursued by system "A" would be as follows: let "Y" represent "A" and its present environment, and let "X" equal "A" in a state in which it has reached its goal. Then "A" will show goal-directed behavior if it performs such internal or external movement that attempts to *minimize* the difference between "X" and "Y." This definition of goal-directed behavior is valid not only for living organisms but for artificial cybernetic constructions. It follows from the definition that a system capable of showing goal-directed behavior has to possess certain mechanisms. It is necessary that the system can distinguish between states "X" and "Y"; that is, it has to have some kind of *recognition subsystem*. Furthermore, it must be able to change its own state, based on this recognition process, which in the case of artificial systems can be very simple (e.g., the heat sensor of a thermostat). Finally, it is necessary that the system carry *internal representations* of possible goals, for without them it could not recognize differences and change.

Subsystems of recognition and change are well known in living organisms. Inner representations are made possible by the mechanisms of memory. However, the question of the organization of internal representation is more complex. Physiological adaptation is principally a relationship to the environment. An organism's adaptive "goal" is to survive in its environment, so the internal representation must be one of the environment including the given organism.

The environment's internal representation can be considered as a construction, which in essence is a model—more precisely, a dynamic model of the environment. By model we mean a systems science definition; i.e., a model is always a simpler system in which the components and the interactions of components *reflect* the components and some interactions of a more complex system (Mesarovic 1964). Model-building, therefore, is always a kind of simplification and a special identification between two different systems, of which one is the model and the other is the system being modeled. The model is used by being operated, and based upon its operation we predict the behavior of the system being modeled.

The environment is an enormously complex system, and its simplified model is a representation constructed by the animal brain. This model includes environmental factors and those interactions most important for the survival and reproduction of animals.

The animal brain's most important biological function is *the construction of a dynamic model of the environment, the continuous maintenance and operation of the model, and use of the data obtained by operation for predictions in the interest of the animal's survival and reproduction*.

If we consider cognition (as MacKay has done) as the ability to construct models, a very useful concept is obtained. It is obvious that all animals, including humans, belong to a common class, because all nervous systems are able to model their environment. That is, all animals have some kind of cognition. This definition reflects not only similarities but differences, because a complex system can be modeled both very simply and in a more complicated way. Simple models also predict some simple but important things for adaptation. For example, each animal "knows" that dark periods are followed by daylight. It is the *model* of the dark-light rhythm of the environment that establishes the diurnal cycle in the nervous system of animals. More complex phenomena also can be modeled. The brain of a wolf can make a model of not only the diurnal cycle, but the behavior of various prey animals, the responses of fellow members of the pack, the order of dominance within the pack, the effects of past events, etc. Ethological studies show that higher mental functions also can be found in animals (Griffin 1976, 1984, Menzel and Johnson 1976, Denett 1983, Epstein et al. 1984), as also has been demonstrated by electrophysiological experiments (John 1972).

As long as the interneuron network is of low complexity, the modeling of the external world also is extremely primitive. The model's structure consists of a few "excited" elements and their simple, temporal connections. In higher organisms strikingly exact images of the external world can be found. Fish, amphibia, birds, and mammals are all capable of storing the essential parameters of an encountered image in their memory and also of acting expediently in a later situation without the presence of relevant stimuli (Beritashvili 1971). Rats can memorize maps of complex mazes, and, by comparing the actual stimuli with the internalized map, they precisely can orient themselves in physical space (O'Keefe and Nadel 1974, Olton and Samuelson 1976).

According to MacKay, the neural model is not only a simple projection but a kind of complex reconstruction containing *instructions of the organism's possible behavior* in response to the external world's stimuli (MacKay 1951–52, 1965). The animal's activity is not organized simply by responses to external stimuli, but also by expectations and analyses of situations based on internal analysis of the model formed in the nervous system (Gallistel 1980). The model is a dynamic representation of the animal's environment—including sequences of events—that the brain uses as an internal reference for control in eliciting fear (Hebb 1946), orientation (Sokolov 1960), attack and defense (Archer 1976), avoidance of predators (Csányi 1985a, 1985b, 1985c), etc.

In higher animals the formation of an environmental model to a great extent involves the internal representation of the animal itself. This process has culminated in the emergence of consciousness in apes (Gallup 1970) and man (Anderson 1984). Interestingly enough, this hypothesis, having appeared very early in behavioral science (Krechevsky 1932, Tolman 1932), was overshadowed by various trends of behaviorism. Recently, ethological studies have provided further evidence of the existence of internal models and images and even of their important role in animal behavior—for example, in the formation of prey-predator relationships and in the appearance of genetic polymorphisms (Tinbergen 1960, Croze 1970). A brilliant systems science analysis of brain functions was given by Lászlo (1969).

The structure of the modeling brain

In getting to know the various neural structures we proceed naturally from the simple toward the complex. Until recent years the amazing complexity of the cortex, at the top of the hierarchy, seemed to pose an almost insolvable problem for neurobiologists. The first hope to find an organizing principle was based on experiments indicating that in the cortex (especially in sensory areas) there are discrete structural units (Mountcastle 1967, Woolsey and Van der Loos 1970, Hubel and Wiesel 1974). Szentágothai's module concept was a breakthrough; its elaboration yielded direct anatomical evidence, too (Szentágothai 1967, 1972, 1975, 1978a, 1978c). Parallel cylindrical structures of a diameter of 200 to 300 m can be detected in the cortex's complex neural network. These structures have ample connections both with other subcortical structures and with one another. Each column contains several thousand neurons, and their internal structure is strictly determined. Between adjacent columns there is a mutual synaptic connection of a "quasi-random" character extending to a more than ten-column distance. Szentágothai regards the anatomical structures corresponding to the column as a kind of *module* analogous to integrated electric circuits, and playing a basic role in the cortex's physiological functioning. The first brain theories based on the activity of the module units already have been set forth (Eccles 1976). Perhaps the most promising is a kind of combination of the module concept and Edelman's "group selection" theory (Szentágothai 1978b).

Edelman's phenomenological model assumes that there are "groups" consisting of a few hundred to a few thousand neurons capable of signal pattern recognition. The signal patterns arriving at the cortex seemingly select from the "repertoire" constituted by the groups. The groups are degenerate; that is, the same pattern can be recognized by more than one group, depending on their actual state. The "primary" groups develop in the course of embryogenesis, and their internal connections are largely determined genetically. The degeneration of the groups is the basis of associative recognition. Edelman's second basic assumption is that the signal patterns reaching the cortex repeatedly "reenter" after the first group selection and in each cycle make a new selection among the primary groups. As a consequence of degeneration, certain groups are exchanged for others when new signal patterns enter, while other groups may repeatedly get excited. The groups frequently excited undergo permanent synaptic changes and compose a "secondary repertoire," which is the basis of memory. Edelman (1977) believes that group-degenerative selection and the repeated reentry of signal patterns provide the basis for higher mental processes. Although a phenomenological model is not supposed to account for underlying mechanisms, there is an obvious analogy between the "groups" of Edelman's primary and secondary repertoires and Szentágothai's modules, as was noted by Szentágothai (1978b). Thus one of the basic assumptions has been put on a firm footing. The postulate of the reentry of the signal patterns by Edelman also seems to be fairly realistic.

The first notion which holds that a critical inner process sensitive to external effects plays a role in the storage and permanent processing of information entering the brain is almost eighty years old. Based on examinations of partial memory loss in traumatized patients, Müller and Pilzecker proposed a "perseverative" neural process, supposedly playing a part in the storage of long-term memory (Müller and Pilzecker 1900). They found that after serious accidents patients are unable to remember events preceding the accident by one to two minutes, probably because the time is too short for forming permanent memory traces. Similar data also have been accumulated (Russel and Nathan 1946). It is supposed that the neural basis of perseverative phenomena is provided by "reverberative loops" formed in the neural networks (Hilgard and Marguis 1940, Hebb 1946), the existence of which could be directly shown (Burns 1958). Thus memory has an unstable phase, based solely on the electric activity of neurons. Recent studies have disclosed further and finer details. First of all, they have shown that smaller gradual electric changes of potential also can propagate in a neural network, especially through dendritic connections (Schmitt et al. 1976). The effect of a weak electric field on the biochemical activity of the membrane of neurons was successfully demonstrated (Bawin and Adey 1976). Besides the well-known "projective" neurons with long axons, a decisive role in shaping the nervous system's specific features is attributed to the "local networks" containing only a few neurons. These microelectric circuits are to an even greater extent sensitive to any change of electric potential (Rakic 1975). There are groups consisting of a higher number of neurons showing a mosaic of coordinated temporal electric activity that is constantly transmitted to distant areas of the brain. The activity mosaics are elicited not only by external stimuli but are connected with the recollection of memory. It is highly probable that consolidation of memory is based on the repetition several thousand times of the temporal activity of neuronal groups (John 1967, 1972). Presently, the exact physiological function of modules still has not been clarified; neither have the mechanisms of the formation of rhythmic activity patterns observed by John and others. It is very likely that the modules play a basic role in processing information getting into the cerebral cortex, and some causal correlation probably will be found between the two phenomena. The best phenomenological approximation at present is still represented by Edelman's model.

The modeling activity of the brain manifests itself in the development of *higher organization* above the level of neurons. Milner (1976, 1977) assumed the emergence of structures consisting of intercellular connections that would correspond to the stimulus, the drive, the response, or even the expectable results of possible responses. These associative structures are *concepts*. The confluence of concepts forms higher active neural networks that determine the animal's behavior in a given situation. These neural networks represent the brain's modeling activity. Neurophysiological experiments were carried out to prove the existence of concepts (Stuss and Picton 1978). The connections between motor functions and cortical models were examined in detail, and the connections between cortical areas and motor activity were found to be quite loose (Pribram 1976). It seems that processes taking place in the cortical regions are not so much linked to motor activity itself as to the purpose and consequences of motor action (MacKay 1966, Evarts 1967, Grastyán et al. 1978). Watanabe (1979) found certain groups of neurons in the cerebral cortex of cats that showed an activity correlated only with the context and content of the stimulus and having no connection with its physical parameters. Therefore, it is certain that the models produced in the brain somehow can be converted into motor activity. A basically similar situation was found in the control of the neural mechanism of speech. It can be demonstrated that cerebral representation of each concept is separately activated and that representation of the corresponding word is activated only at the time of verbalization, which in turn elicits the motor patterns of the larynx during speech (Penfield and Roberts 1966, Johnson-Laird et al. 1974).

Study of the brain's modeling activity is one of the most rapidly developing fields of neurobiology, but it is beyond the scope of this work. There still are two important things to be mentioned here, however. First, the modeling process is not a kind of photographic representation, but rather a complex, hierarchically ordered abstract organization of information. The mass of information entering the brain from the peripheries is concentrated and interpreted by the brain, and only selected parts (or excerpts) pass to the next levels of hierarchy where concentration is repeated several times (Jerison 1978). Second, by computerized electroencephalography it is possible to analyze data of the cerebral processes connected with sensation, subjective feelings, and conscious activities. There is a detectable difference in EEG waves, for example, when in a written text someone recognizes the same word as a noun or verb. Also it can be recognized that someone is expecting a special, external stimulus. There is a detectable invariant component, for example, in the recognition of squares of various sizes; that is, the abstraction process can be traced. Special memory content can be recognized, as can the recollection of memory and its comparison with present circumstances, etc. (John et al. 1977).

We should point out that although the purpose of most brain models is to explain human cerebral activity, it is almost generally accepted that, concerning the basic processes, the brains of vertebrates and among them those of mammals are only marginally different from one another. It also is obvious that the higher processes, e.g., conscious activity, have developed not by leaps but gradually, and that differences between individual animal species are more of a quantitative character (Griffin 1976, Menzel and Johnson 1976, Sperry 1976).

Replicative memory model of the brain

Accepting the Szentágothai-Edelman-John-Milner conceptions, we should point out a completely novel aspect of the relationships between various phenomena: notably, that in the brains of vertebrates an autogenetic system is working. The evidence concerning this idea will now be summarized in our new brain theory.

Excitable building blocks of the autogenetic model can be identified with neurons, and the higher structures formed by the interactions of neurons are considered *components* of the neural system that are continuously built up and dismantled. The inner structure of these components is still uncertain.

The brain's modeling activity manifests itself in the formation of higher organizations. How are these organizations constructed? According to Edelman and Milner, they consist of neurons and neuronal connections. According to John, they are composed of temporal patterns of electric potential changes. There also is a third possibility: the organization might be based in the temporary electric connections of the Szentágothai modules. At present, it would be fairly difficult to adopt any of these possibilities, and a new theoretical solution—or some synthesis of the existing theories—cannot be excluded. The John hypothesis seems the most plausible because it allows the greatest freedom for structures to be formed. The same potential pattern can develop in various parts of the brain; that is, the same structure can be transferred to the most different parts of the brain. Because of the differences among fixed neurons, the other two hypotheses would allow only a much more rigid and limited superstructure-building.

The self-organizing properties of neuronal groups were analyzed on a theoretical ground by Dalenoort (1982).

This question, however, can be left open. There is sufficient evidence for the existence of some kind of "superstructures" in the brain that developed as a result of neuronal activity. The actual form of these structures is not fundamental for our present purposes. During the development of the gene concept there was a similar situation in genetics. For a long time the gene was believed to be a complex protein structure, and the foundations of genetics were successfully laid without revealing the true nature of genes.

Let us follow our line of thought by stating that brain activity is accompanied by the formation of *organized structures* that we regard as *components* of the neural system. After Milner, we call these components *concepts*. Their appearance and state of excitation obviously depend on already existing states of excitation. Namely, the concepts can influence the *probability of genesis* of each other; i.e., they have functions in terms of definitions used in connection with the autogenetic theory. The evolutionary zero-system characteristically contains excitable units that form higher structures (components) having functions. Therefore, it is worth examining whether other characteristics of evolutionary phenomena can be found in brain activity.

Concepts as replicative structures

If concepts are assumed to develop by the effects of external stimuli through group-degenerative selection of neurons and reentrant signals, essentially a replicative process is supposed. Each cycle of the reentrant signal mechanism corresponds to a replicative cycle. The rhythmic temporal activity pattern of large neuron groups also means multiple repetition and also could correspond to a series of *replicative cycles*. The question is the fate of the replicates produced. It is certain that in most cases they do not disappear without trace since the "secondary repertoire," i.e., the lasting memory, develops just as a result of the reverberation of repeated entries. It means that concepts are in two states: in their "resting phase" they exist in the form of lasting memory traces; in the "multiplicating phase" they go through thousands of replications in the course of repeated reentries, and finally they get into the resting phase again up to the next recollection. The problems of recollection cannot be dealt with here, but independent of whether concepts are made of neuronal connections or potential patterns it should by all means be assumed that these structures can be repeatedly "recollected." John and Edelman in their model also offer some ideas concerning this problem (John 1967, Edelman 1977). During replication of concepts each "copy" corresponds to the memory trace developing in one reverberating cycle. Since on the average several thousands of replicative cycles should be reckoned with (John 1967), the number of copies of evolving concepts can be quite large, and probably they are stored most economically. Unfortunately, not much is known about the mode of storage of lasting memory traces. The majority of authors believe they are stored not in well-defined, discrete neuron groups but in some kind of "dispersed" storage mechanism (Lashley 1950, John 1972, Pribram et al. 1974). If, as we assume in the replicative cycle, concepts can spread over large zones of the brain, then in the resting phase they can be found in *many copies* and over relatively large regions.

In this case, however, the phenomena providing the basis for various hypotheses of "dispersed" storage can be interpreted even assuming a completely discrete storing mechanism. For example, Lashley's (1950) "law of mass action" could be seen as a direct consequence of the replicative model. The "law of mass action" was based on experiments in which cortical areas of various sizes had been surgically removed in experimental animals before their problem-solving ability was tested. It was found that by removing increasing pieces of the cortex, the animals' achievements became more and more superficial. But deficiency in any special feature of the task was not detected. This is very difficult to interpret unless we assume the simultaneous and parallel presence of numerous copies of the concept structures in the "module field," all having nearly the same function and effect. This may happen, for example, when through reentrant signals the excited neuron group activates the part of the network that is *similar to itself*, or when the patterns of the activated part of the neuron network are slowly "spreading," always activating newer parts of the "module field" while always corresponding to the original concept structure in its pattern. If, for example, it is assumed that a single module element is capable of recognizing various signal patterns, but on entry of each pattern it "assigns" only a single neuron or some specific neurons as elements of the developing concept structure (that is, the deposit of lasting memory), then the same module can be part of various concepts, and the same concept can exist in many parallel copies. The rhythmic activity of the simultaneously activated concept population can be seen as Szentágothai's "dynamic patterns" (Szentágothai 1978b). These questions, however, can be decided only by becoming more familiar with the physiological functions of modules and with the mechanism of long-term memory.

Interestingly, theoretical questions concerning the replication of structures consisting of excitable elements have already been dealt with by several workers. Von Neumann (1966) devised the general theory of self-reproducing automata. The basis of his two-dimensional model is a planar quadratic lattice in each cell of which there is an excitable finite automaton of twenty-nine possible states. Its actual state is determined unequivocally by the state of adjacent cells. From about 200,000 cells a self-reproducing structure with universal computing and constructing abilities can be built. Since Neumann's investigations, numerous simpler self-reproducing cellular automata have been devised (Codd 1968, Burks 1970, Laing 1977).

As far as the basic principle is concerned, there is considerable similarity between the replication of concepts consisting of the excitable Szentágothai modules and the functioning of the Neumann cell-automaton. The possible number of states of the real module is definitely over twenty-nine; the real superstructure is not two- but three-dimensional; and perhaps for this reason fewer than 200,000 elementary units are sufficient to construct one replicative structure. The crucial question is whether the activated structure really undergoes *physical replication*, which is the basic postulate of the autogenetic model.

The question of the concept structure's stability has not been dealt with here. However, there are two alternatives. The life span of individual concepts in the resting phase can be in the range of the organism's life span; that is, the decay of concepts during storage need not be reckoned with. A fairly great number of human studies support this possibility (Penfield and Roberts 1966); so do a few data on animals (Skinner 1968). The second alternative supported mainly by experiments on animal memory (Wickelgren 1972) would be that the concept components have a much shorter life span than the organism. Concerning our model, both alternatives have about the same value, although each requires specific supplementation that will not be discussed here.

Selection, mutation, and recombination of concepts

The idea of some kind of selective mechanism in mental processes already has been proposed on an intuitive basis (Pringle 1951, Pattee 1965, Darlington 1972, Campbell 1974, Glassman 1977). Dawkins (1976) even has assumed the multiplication of memory traces analogous to the multiplication of genes. His ingenious line of reasoning about the struggle for survival of "memes" is very close to our theory, but as far as we know concrete models have not been worked out.

An inherent feature of the replicative memory model is the assumption of selection. The replication of concepts takes place under the continuous controlling effect of various motivational mechanisms and the environment. Concepts having adaptive value for an animal get into the multiplication cycle more often, and therefore they are produced in greater numbers than irrelevant ones or those that are harmful. As the number of modules is limited, the concepts are in constant competition for excitable module elements. Differential reproduction of concepts, being the essence of selection, is unambiguously realized. It is easy to see that "mutational" changes of concepts also occur. Selection of elements of the concept component is a group-degenerative process; that is, it often happens that in a replicative cycle an element with new properties is incorporated into the concept. The competition of different variants directly follows from the selective mechanism. It is conceivable that recombination of simultaneously replicating concepts also may occur, thus producing new, even more adaptive forms.

Autogenesis at the neural level

In the previous discussion the existence of replicative structures in the bra·n has been examined; now we will discuss whether the autogenetic model suitably describes their behavior.

We might begin by examining the conditions necessary for spontaneous autogenesis in the nervous system during evolution. It is generally accepted that the nervous system of simpler animals is largely "wired in"; that is, a majority of neuronal connections are genetically determined. At the same time even the most primitive organisms show habituation and sensitization, providing a possibility for the emergence of simple concepts. The study of these primitive animals certainly would provide important data, but it also seems essential to examine whether highly developed nervous systems are predictable on the basis of our autogenetic model. The most important questions are

(a) Can traces of replicative organization be found in the nervous system? (b) Are there any system precursors?

So far we have found that replicative organization emerges when the dynamics of the components of a given organizational level are influenced and controlled by *constraints* that can be interpreted on a higher level. In the case of the nervous system this higher level is behavior. For an analysis of the nervous system's organization, data must be provided by the behavioral sciences. Behavior is the mode by which the animal secures the integrity, maintenance, and reproduction of its organism. Behavior is the output of the brain, and at the same time it is the "interface" that relays the effects of the environment to the animal. The animal's body structure and its niche represent possibilities and constraints that also influence the animal's brain. These constraints determine the boundary conditions under which the animal's behavior is optimal and adaptive regarding survival and reproduction. However, this does not mean that all behavioral patterns are a priori *adaptive;* we must search for these, and solid evidence for their adaptive value must be provided. If the environmental model of the brain built up from concept components has indeed a replicative organization, then we must find the factors responsible for this at the level of behavior.

Outlines of a behavior theory

There are three important characteristics of animal behavior: it is cyclic, determined largely by genetical factors, and modifiable by environmental influences.

(a) Behavior is cyclic. Maintenance of life and reproduction are based on hierarchically organized cyclic processes (figure 17). In ad-

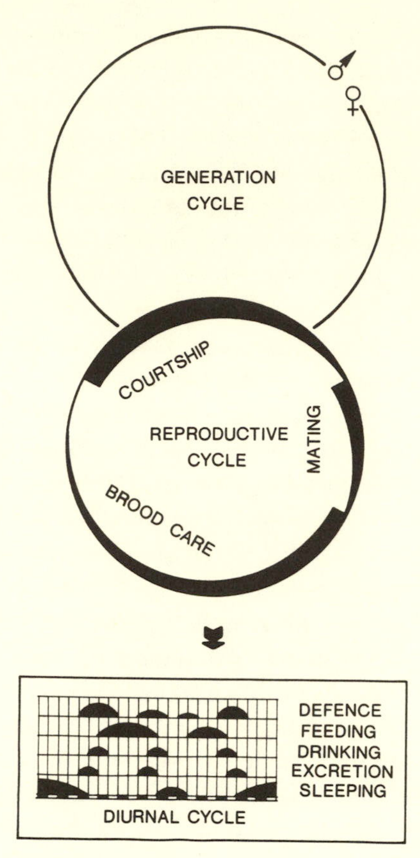

Figure 17 Various cycles of animal life.

dition to daily feeding and resting cycles, there are seasonal cycles of reproduction and migration. Encompassing these is the great cycle of changing generations. This cyclic organization is the basis for the type of cyclic activity in which *concepts generate behavior units*. Cyclicity is the condition that provides the mechanism for changing concepts, ensuring their recurrent activity. This cyclic organization is the exact analogue of the reproductive cycles of a molecular autogenetic system and of the replicative cycles of cellular and organismic autogenesis. Concepts not capable of cyclic activity will be selected against and will not participate in a replicative organization. All other concepts can exist, provided their cyclic activation and the resulting behavior do not interrupt the great cycle necessary for survival and reproduction. *Replicative selection* of concept components is provided by this organization. It is a creative process since all concepts that are not harmful in the above sense can survive whatever behavior they activate. Practically, their survival also depends on stabilizing selection in two senses. There is competition among concepts. Reinforcement mechanisms of the brain favor concepts that lead to adaptive behavior in the given environment. On the other hand, in the case of strong competition among individuals, those individuals that can generate better (more adaptive) concepts will survive. This latter selection type influences the animal brain's concept-generating mechanisms.

(b) There are genetically determined behavior patterns that are expressions of genetically determined concepts. Fixed action patterns, reflexes, innate preferences, and aversions are all forms of behavior that are maintained by such genetically determined concepts (Csányi and Gervai 1986, Csányi et al. 1985a, 1985b).

(c) Another characteristic of behavior is that it can be modified by the environment. One way, among others, it can be modified is through *learning*. The idea that there is a close analogy between the learning process and biological evolution is not new, and it appears frequently in studies of animal behavior (Pringle 1951, Skinner 1966, Staddon and Simmelhag 1971, Pask 1972, Cloak 1975). However, so far no detailed theory has been elaborated that by its predictive power could supersede existing theories. Without being able to set forth such a theory, I would like to point out a few basic questions that arise while devising such a theory.

A comprehensive learning theory should meet at least the following criteria:

It should explain the mechanism of phylogenetic and ontogenetic variability of behavior.

It should give a unified explanation of the neural organization of inherited and acquired forms of behavior, integrating the "ethological" and "neurophysiological" approaches to the study of animal behavior.

It should give a clear neurobiological explanation of the cyclic nature of various behavioral forms.

Variability of the behavioral phenotype

Behavior, like any other phenotypic character, can be traced back to the animal's genetic constitution, and certain forms of behavior are remarkably variable. Two distinctive factors create this variability. One is the varied genetic constitution. The other is the learning process. As phylogenetic selection presupposes the existence of genetic variability, the existence of *genetic variations of behavior* obviously follows. A wide range of genetic variability of behavior is known in behavior genetics (Fuller and Thompson 1978, Fuller and Simmel 1983, Csányi 1985a, 1985b, 1985c; Csányi and Gervai 1986, Gervai and Csányi 1985). On the other hand, the biological function of different types of learning is to provide an even greater "*ontogenetic*" variability above the level of genetic organization. The main selective factor of the sequence of behavioral patterns developing through learning affects the nervous system and shapes the most adaptive forms of behavior in a given situation. The existence of this selective mechanism has been elaborately analyzed in the study by Staddon and Simmelhag (1971).

Concept: a functional unit of behavioral regulation

It has been concluded that the activity of the animal brain manifests itself in producing and maintaining a kind of environmental model. Such a model may correspond to the activity of a single concept or several joint ones. If concepts are replicative, as has been assumed, then it is selection operating in the replicative process that acts as the main factor in creating ontogenetic variability of behavior. *The essential process of learning is the selection of concepts generated by the brain.*

How is the genetic variability of behavior brought about? To answer, the concept structure has to be further dissected. The concept, as a neural model, is assumed to be composed of smaller, genetically determined "elementary" behavior programs called *instructions* (MacKay 1951–52). These are necessarily unchanged during a lifetime, but as a prerequisite of their phylogenetic development they must display adequate genetic variability within the population. As Cloak (1975) has suggested, it is expedient to suppose that elementary instructions are generally connected with an environmental stimulus cue, and the

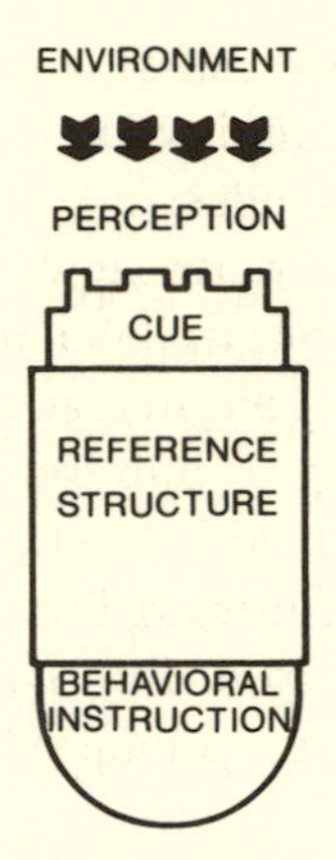

Figure 18 Building blocks of the environmental model of the brain.

appearance of a definite stimulus cue activates an elementary instruction. If we want to include learning phenomena as well, this mechanism has to be supplemented. As has been noted, even the simplest of nervous systems are characterized by a triple division. Separate receptor neurons interact with stimuli; motor neurons create or mediate behavior instruction; and interneurons, between receptors and motor neurons, serve as a *reference* for the activity of the three as a whole. The actual state of the network of interneurons, or their past activity, influences the choice of the actual behavior instruction and the extent of response (if response occurs at all). The phenomena of sensitization and habituation in simpler animals are good examples. These three functional components of behavior are retained even by the higher nervous systems, but the complexity of parts increases enormously and functional overlaps have developed.

A higher animal is not "stimulated" by a stimulus but finds itself in a "situation." It searches for food; it is chased or chases others, etc. Effects of light, sound, and mechanical vibrations signal concrete situations such as the presence of a predator or prey. Stimulus components of the particular situations elicit no isolated responses. More precisely, not only reflexes or fixed action patterns are activated, but the animal is in *action;* it escapes, attacks, or is thinking what to do. The reference is not simply a state of some neurons; but memory traces and their mutual interactions create experience and a general view of the environment. This operating scheme is shown in figure 18.

Perception cue

The sensory apparatus is mainly genetically determined. Sensory organs of higher animals do not simply perceive and measure one-dimensional stimuli, but there are complex "interpreting" mechanisms that mediate *relations* of the external world to the central nervous system. The retina of the frog's eye, for example, is itself a complex nervous system that extracts important and expedient information from the environment. The retinal perception of higher animals like cats or monkeys contributes less to the processing of information, primarily mediating changes and contrasts. The interpretation of these phenomena occurs at the brain's higher centers (Arbib 1972). It is not farfetched to assume that information processed by perception is divided into elementary units—for example, the textrons of Julesz (1981) or the perception of linear or oblique contours. Key stimuli defined by ethologists also belong to this class (Lorenz 1981). Genetic determination prevails in these small units, while the higher construction built up from them is learned. Put simply, perception of contours and lines is inherited; that of the cube is learned.

The ontogenetic influence of perception was shown by the famous experiments of Hubel and Wiesel (1974). They found that line perception in young cats is bound to the activity of certain groups of neurons, and these are able to develop normally only if the young animal receives appropriate visual stimulation. If such stimulation does not occur at a certain early age, then the animals later are unable to perceive horizontal or vertical lines.

Learning also influences perception. Specific "search images"—that is, when the animals are looking for specific food, prey, etc.—are clear examples. "Innate releasing mechanisms" also may be considered perceptions (Lorenz 1981). Such releasers (the key stimuli) were found to play a role in parental behavior in birds. A certain configuration of stimuli, including several key stimuli, may induce adaptive responses without any previous experience. It can be shown that it is not the complete configuration of stimuli that influences the animal's behavior but the key stimuli acting separately; the behavior is a sum of their effects. After several repetitions additional learning forms a kind of "Gestalt" from the participating stimuli and the situation's other features. Releasing mechanisms are relatively primitive forms of response. This is so because stimulus and response are not yet separated, and perception is immediately followed by action without previous consultation with the reference neurons. Relative simplicity also is shown by the fact that releasing mechanisms play a dominant role in lower animals such as insects, other invertebrates, fish, and birds. Releasing mechanisms are occasionally found in higher animals, but very often learning may modify their effects. There are some properties of the physical environment—namely, space, objects, light, etc.—that are equally important for all higher species. Visual perceptual units such as lines, contours, angles, and movement are generally perceived by animals. The key stimulus-releasing mechanism complex has a similar informative function, but the message is *species-specific*. This complex mediates information from objects such as parents, progeny, food, etc., relevant to the given species. Just as in the visual perception process, only the most essential part of the information in the environment is represented by the key stimulus.

Referential mechanisms

Environmental information reaches the central nervous system after being processed by receptors, and this processing continues in the higher centers. All mechanisms located in the central nervous system and dealing with further interpretation and processing of information are regarded as *referential mechanisms*. A part of these referential mechanisms is genetically determined. Constraints resulting from the animal's morphology, its primary motivational systems, inherited fears, phobias, and preferences, mechanisms of the internal rhythms, hormones—all belong here (Hinde and Stevenson-Hinde 1973).

Another part of referential mechanisms is formed during the ontogenetic process; learned ideas, plans, intentions, and especially the *memory* belong here (Miller et al. 1960). To unite them, we employ the concept of *cognitive space*, often used in cognitive psychology. The animal brain's cognitive space is an abstract space embodying all mental factors that participate in the processing of information mediated by perception until the decision to select the relevant action is taken. Cognitive space has a very special structure. Traits of given species assign certain sensitive points or fields, and memory traces bound to these play special roles in the animal's life. In other parts of the cognitive space, memory traces can be built up only with great difficulty or not at all; i.e., the animal is unable to fix certain associations. This may be illustrated by the concept of "preparedness" used in the physiology of learning. Rats are unable to associate the shape of food with its toxic effects. However, association is very quick (one trial is enough) between the taste of the food and its toxicity. It is just the opposite in some other species. Quails, for example, are able to associate color and shape with toxicity but unable to do so between taste and toxicity (Garcia et al. 1966). Cognitive space is nothing like a Euclidean space that has homogeneous parameters in the whole universe of space. Cognitive space is multidimensional; its parameters change as a function of position in the space. The more developed a species, the greater are its possibilities to position a memory trace in its cognitive space. But this placing process depends not only on the number of repetitions but on the genetically determined parameters of the cognitive space. Genetic and environmental effects expressed in the cognitive space are mutually complementary. Without environmental effects (as in the case of an isolated, deprived animal), the cognitive space remains almost empty because only the actions of an active animal and responses of the environment together can build various concept structures into this space. But the rules of this construction are determined genetically. We put a greater emphasis on genetic constraints because the behaviorist tradition of psychology has done just the opposite, having attributed importance exclusively to learning. Learning has been considered as a general process, but recent results urge us to change the old slogan that "everything can be taught and everybody can learn" to "everything can be taught to somebody, and everybody can learn something."

Properties of the cognitive space are important not only with respect to genetic constraints of the memory construction processes. Higher animals and man have a very complex cognitive space, and the concept structures developed in them influence the probability of one another's genesis. Thus a higher organization—a new superstructure—is created (the mental "personality" of the animal), which varies according to the sequence of acquisition of new information, thus producing highly varied individual behavior.

An important parameter of cognitive space is that its structures can be *transformed* according to certain rules. Memory traces can be combined, and new solutions to problems may emerge. These transformations are active processes and are based on the fact that the cognitive space can accept information not only from outside but from internal processes. When the brain prepares a plan, the external information simply initiates the process, and future action is built up by transformation from structures already existing in the cognitive space. Thus there is an active *internal construction mechanism* in the cognitive space making the behavior of the higher animals so varied, planned, and conscious. Summing up, the cognitive space of the animal brain is a complex, dynamic system embodying hierarchically organized, mutually interacting structures. It can manifest itself as a systemic unity, a kind of "worldview," but of course not in human terms. This cognitive space governs the animal's general behavior and its relations to the environment.

Behavioral instruction and action

Finally, we should analyze the third functional unit of behavior: behavioral instruction. As we have noted, all animals possess numerous genetically determined behavioral instructions—responses such as reflexes, taxes, fixed action patterns, innate aversions, and preferences. These responses are closely connected to environmental stimuli. If an object approaches the eyes of a man or animal, they respond with a defensive reaction, with a reflex, and there is no need to consult referential mechanisms. In the same way releasers and key stimuli directly activate various inherited behavior patterns of the lower animals. There is a tendency in evolution for the generalization of responses. That is, components not directly connected with the primary stimulus-response unit may contribute to the activation of a response. Fixed-action patterns can be elicited or inhibited by conditioning. Males of the jungle fowl cannot be conditioned to perform their courtship display for food rewards, but they readily perform the required patterns when given access to a female as a reinforcement (Kruijt 1964). This is a mental transformation; the features of the female can be substituted temporarily by the features of the closed door of the conditioning apparatus. The mental rigidity of this species is shown by the males' inability to associate the courtship dance with a food reward; in other words, they are unable to perform this mental transformation.

During evolution not only the possibility of neural transformations was created; in the phenomena of ritualization (Eibl-Eibesfeldt 1970) stimulus control of inherited elements of behavior genetically change and become controlled by a completely different group of stimuli. Also genetically determined is the feedback mechanism called the TOTE (Test-Operate-Test-Exit) unit (Miller et al. 1960), which makes goal-directed behavior possible for the animal.

Inherited responses are often complemented with learned instructions. Animals are able to store engrams of complex behavior sequences in their cognitive spaces, and in appropriate conditions these constructions direct their behavior. Evidence was found showing the existence of internal representations of learned motor responses in both man and animals (Terzuolo and Viviani 1979). The acquisition of skills is based on these representations. These fixed motor sequences may be very rigid and expressed in the form of stereotypes, or they may be somewhat flexible and subject to internal transformations. These learned motor patterns have an important role in animals' various behaviors. It has been proved beyond doubt that sometimes an animal's response is much faster than would be possible if it used feedback mechanisms of the proprioceptor structures. That is, the brain gives orders in advance necessary for the next future motor response. When a well-known object, an apple, for example, is offered to a monkey or a man, upon taking it they usually do not drop it because the muscles of the arms use the amount of force needed to keep an apple in hand. We might think that the brain received some information about the apple's weight through the proprioceptors located in the arm and that this information served as a control in maintaining the right muscle tone. However, this is not the case. When the experimental subject received a "prepared" apple containing a hidden weight, e.g., a piece of lead, the arm was unable to keep the apple in the right balance; both the apple and the hand moved downward, and sometimes the apple was dropped as the *unexpected* weight remained uncompensated for. As the arm moved down, the appropriate feedback began to work, and with some delay muscles assumed their correct tones. Electrophysiological experiments showed that when an

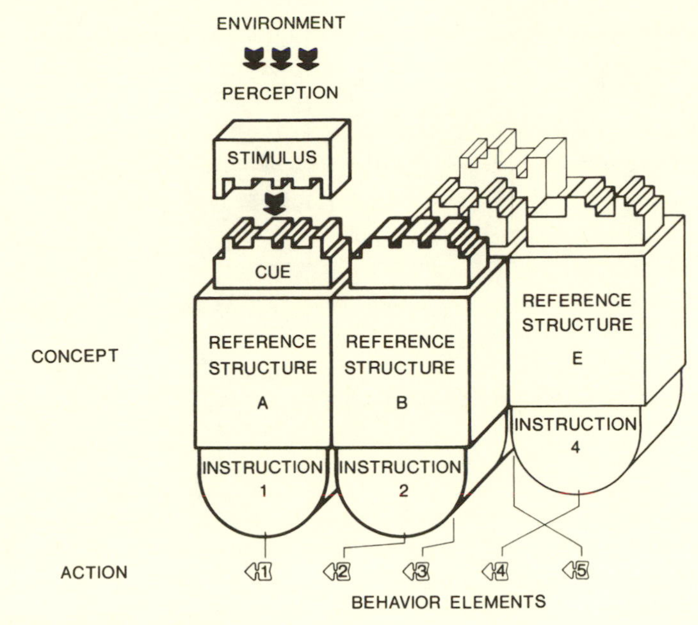

Figure 19 Structure of a concept.

apple is offered to monkeys, the brain gives its calculated order several ten milliseconds in advance, and the feedback mechanism makes only later corrections if necessary.

In the case of more primitive animals these mechanisms are simpler, and responses depend on the complexity of the whole cognitive space. So far we have been speaking of responses, but an isolated response is very rare. Animals perform highly complex *actions*. In a laboratory environment an animal does not simply respond to the tone of a bell activated by the experimenter, but it wants to go out from a cage, or if hungry it wants food. That is, the animal responds to the stimuli through actions. The action is an *organized response* built up from units of simple responses.

The concepts discussed so far are illustrated in figure 19. The concept component is the higher structural element of the brain's neural networks; a set of the three units: *cue–referential mechanism–instruction*.

Finally, we want to emphasize that concepts are not static but dynamic structures that work in parallel modes (Arbib 1972) and par-

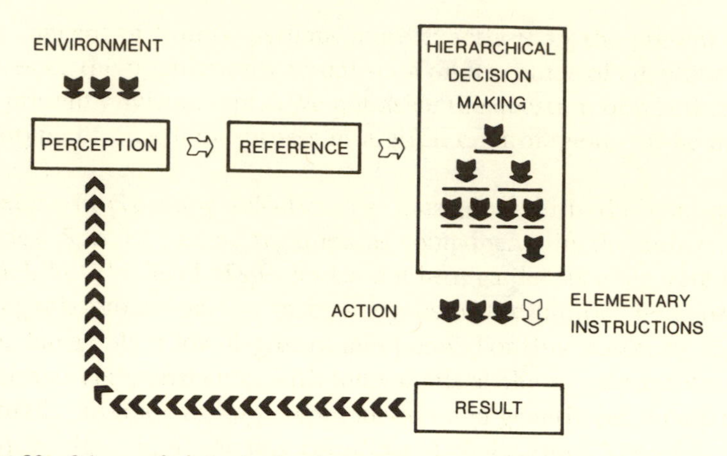

Figure 20 Scheme of a behavioral action.

ticipate in hierarchically organized decision mechanisms (Dawkins 1974) that are able to choose among many possible responses and many possible subroutines (Gould and Gould 1982). The state of the environment may change from moment to moment, and actions are either successful or not; the result of actions is immediately represented in the brain's cognitive space and influences further events (figure 20).

In summary we want to emphasize three factors: (1) The animal brain is a *constructing machine*. Its constructions are the *concepts* that contain the cue–referential mechanism–instruction constituents created in a replicative system. If we consider the evolution of animals, we can recognize that the animal brain's constructing ability is increasingly sophisticated as we move higher on the evolutionary tree. Even the most primitive nervous systems are capable of recognition, which is basically a replicative structure-constructing activity. As the action repertoire of the animal extends, besides replication, mechanisms of recombination (i.e., transformation) appear. The low fidelity and probabilistic nature of replication make possible the creation of new concepts, "hypotheses." In encounters with its environment the animal's concepts become either strengthened or extinguished in a differential replication process, which is the basis of neural evolution. It is characteristic of this evolutionary process that the supersystem of concepts constitutes a systemic unity (the personality) that has new *holistic properties*.

(2) The replicative constructing activity discussed above is subjected to selective forces connected with adaptation. Concepts are representations (replicas) of parts of the environment, and the *dynamic organizations of concepts are models of the environment* that provide predictions promoting the animal's survival and reproduction.

(3) The animal brain's model-constructing ability is regarded as *consciousness*. It seems well established that animal awareness forms a wide continuum, and its levels in the various species depend on their states of evolutionary development. Man as the most developed animal has broadened this continuum with his special properties, which will be discussed later.

The adaptive value as the basis of behavioral selection

What does influence the selection of individual concepts? Obviously, this is a fundamental question for our theory of behavior. Previous learning theories have been built on the concept of *reinforcement*. It has been postulated that different elements of behavior develop more or less spontaneously, and their repeated appearance can be induced by reinforcement, either punishment or reward. In recent years a number of learning phenomena have become known that could not be interpreted by the simple mechanism of reinforcement. (For details, see an outstanding analysis by Staddon and Simmelhag 1971.)

Here again two levels of behavioral selection should be considered. In the long run selection at the genetic level evaluates the "adaptive value" of a given phenotype. Study of the adaptive value of various behavior patterns that are considered to be inherited is the most important task of ethology (Lorenz 1965). Obviously, the behavioral patterns manifested in the execution of concepts serve to increase the survival of a given individual. From this it also follows that neural mechanisms performing the ontogenetic selection of concepts must, after testing the individual concepts, assess their *adaptive values* and inhibit or decrease the multiplication of those without merit. The same mechanism has to decide the *sequence* of use of behavioral patterns that have already proved to have adaptive value. This selective apparatus is connected with various drive and motivational mechanisms and must be genetically determined. Replicative and competitive selection also can be interpreted on the level of behavior. All concepts harmful to the animal's survival or reproduction are eliminated. Those remaining will be subjected to competitive selection which eliminates concepts that are inadequate in competition with conspecifics. Competitive selection is stabilizing, while replicative selection is creative in nature.

Parametric and replicative information in neural autogenesis

Based on the definitions introduced with the autogenetic model, the parametric and functional information content in neural evolutionary systems can be examined as well. The proportion of parametric and functional information in animals obviously depends on how precise the replication of concepts can be in the nervous system of a given species. Information content in a species capable only of nonidentical replication consists mainly of parametric information, which poorly reflects the animal's individual history. The more accurate the replication, the greater is the increase in the proportion of functional information as well as replicative information during the animal's lifetime. The higher the proportion of functional information, the more accurately the behavior of the animal reflects individual experience. The higher the fidelity of replication, the better the memory preserves past events. An estimation of this proportion, however, will require further experiments.

The next question is whether animal species have reached an evolutionary level in their neural systems that enables them to secure identical replication. The answer is a definite no. The succession of Piaget's developmental phases of behavior (which will be dealt with in detail in the next section) also was studied during the ontogenesis of apes and lower monkeys. It was stated that the six intellectual phases observed in man also develop in apes. Especially in gorillas and chimpanzees, the primary, secondary, and tertiary circular reactions successively appear. However, in the case of lower monkeys the repetitive movements corresponding to circular reactions are missing. Their development reaches only the second of Piaget's phases

(Chevalier-Skolnikoff 1976). In accordance with this observation the apes, or at least gorillas and chimpanzees, have the capability of using signs (Gardner and Gardner 1969, Premack 1971, Lieberman 1972, Patterson 1978). Consequently, these animals have certainly reached the capacity for quasi-identical replication of concepts, but probably they are incapable of true language use (Sebeok and Umiker-Sebeok 1980), which is necessary for the identical replication of concepts (as we will discuss later in the case of humans). With further extrapolation it seems very probable that in the course of their ontogenesis lower animals can reach only the developmental level of nonidentical replication. Their concepts are less differentiated, inaccurately replicated, and fade with time. Dogs and perhaps dolphins are possible exceptions as their abilities in certain respects equal those of apes (Premack 1972). Still, further experiments are needed to prove this hypothesis.

Animals, being in the nonidentical phase of replication of concepts, obviously do not display any phenomena of convergence. These species are mentally "perennially youthful," their modest performance stable and independent of age. The further a species has advanced in the accuracy of replication of concepts, the greater the difference should be between the level of intelligence of young and adult, naturally disregarding differences that come from physiological maturation and aging. Verification of this prediction requires experiments based on a new approach.

Sensitive periods

The ontogenesis of animal behavior is characterized by the appearance of different sensitive periods (Eibl-Eibesfeldt 1970). So far neither psychology nor ethology has been able to provide an explanation for this phenomenon. However, the existence of sensitive periods can be explained without any further assumptions on the basis of the replicative memory model.

The concepts developed in an early phase of neural evolution, i.e., at the beginning of the ontogenesis of the individual, obviously have more favorable conditions in which to replicate than those developing in later phases. Consequently, early concepts will influence the behavior of the animal or man more profoundly than later ones. This presumption is in complete accord with everyday experience. After early phases of life the brain's cognitive space is densely populated with various concept structures, and thus the multiplication of a newly arising concept depends not only on factors directly influencing it (such as drive, reinforcement, etc.) but also on the established state of concept ecology.

Phases of autogenesis of the human brain

Various data suggest that in the course of ontogenesis the human brain goes through phases of neural autogenesis; the phases of nonidentical replication and identical replication and the phenomena of each phase (functional differentiation, rise of supercycles, compartmentalization, convergence) also can be observed. The human brain, which consequently represents a zero-system capable of neural evolution, develops in the course of embryogenesis. The first phase of concept development is one of nonidentical replication. The developing brain can be considered a cognitive space, similar to the animal brain, in which the physical components corresponding to concepts are continuously synthesized. In this synthesis a decisive and guiding role is played by various motivational mechanisms and external stimuli that at a very early stage already have a positive effect on certain types of concepts. The first concepts connected with space, time, motion, and objects develop, and these are considered *system precursors* of neural autogenesis.

There are several reasons to believe that the formation of concept structures follows certain genetically determined internal rules that have developed in the course of the evolution of the species. Linguistic universals proposed by Chomsky (1968) could be such "constructing rules." Preferences or aversions in our thinking manifest themselves especially in social relations. An inherited control system for the development of concepts responsible for these social attitudes might be recognized in the ethnological universals of Lévi-Strauss (1963). There are neuronal networks in lower animals that perform specific, inherited motor patterns (Willows 1967, Sturmwasser, 1973). Analogously, it seems very likely that inherited concepts exist a priori independent of direct experience; these concepts have particularly great adaptive value and are produced by patterns built into the genome in the course of human evolution (Lorenz 1975). The concept of causality fits into this description (Stent 1975). These concepts, either innate or forming very early in life, especially those connected with the "self," begin the replicative cycles that, with further extension, will develop into the concept network of mature thinking. In the early phase of neural autogenesis, concepts develop in relative independence of one another. However, they soon begin to influence the probability of one another's genesis.

The phase of nonidentical replication of neural autogenesis

It might be possible to get an insight into the phase of the development of concepts by studying the intellectual development of children (Piaget 1952). Piaget differentiates between six developmental phases: the first stage is characterized by the use of reflexes; in the second there appear primary circular reactions consisting of "self-directed" repetitive motions; the third and fourth stages are characterized by secondary circular reactions consisting of movements directed toward objects; the fifth is the stage of tertiary reactions in which the child consciously explores objects for new and unknown features; in the sixth stage mental combinations appear as well as the modeling ability of the brain. The very basis of Piaget's classification, *circular reaction*, supports our model remarkably well, as the repetitive motion sequences can be regarded unambiguously as motor manifestations of the development of concepts. Most probably they also can be considered system precursors.

At first the circular reactions are vague, the repetitions are inaccurate, and the development of concepts is clearly going through the *phase of nonidentical replication*. By going through each individual stage, the accuracy of replication rapidly increases. Differentiation can be distinctly observed as well. Piaget's second and third phases differ in the individual's ability to differentiate between the "self" and the external world; i.e., concepts related to the "self" and to objects become gradually separated. The fifth stage is characterized by differentiation of secondary and tertiary circular reactions. According to Piaget, differentiation of secondary circular reactions is "enforced" by the environment, while differentation of tertiary reactions is initiated by the "self." The sixth state is characterized by mental combinative ability that requires a high degree of replicative accuracy in producing concepts; and increasing differentiation indicates the cooperation of concepts, that is, the emergence of supercycles. Each mental model represents a supercycle, and the individual concepts as members of supercycles compose a lasting coreplicating unit through functional connections having developed during the process of comprehension. In later phases of intellectual development there are many more examples of these phenomena (Piaget 1960). In every developmental phase the external stimuli characteristically play an initiative-"inducing" role in the production of concepts.

The phase of identical replication in the neural autogenesis of humans

In humans the phase of identical replication of concepts develops in connection with the use of language. With the aid of language, the very accurate replication of complex concept structures and their storage for as long as a lifetime become possible through adequate memory mechanisms. From data obtained by direct stimulation of the brain it appears that, although conceptual thinking and linguistic formulation are entirely separate processes (Penfield and Roberts 1966, Sperry 1976), intuitively it is easy to comprehend that the two systems are in constant interaction. The grammatical boundaries of language, to a certain extent, limit the structure of concepts, giving higher adaptive value to *linguistically expressible* concepts. On the other hand, the evolution of concepts during an individual life will fill the linguistic "behavior" with a more and more accurate content (i.e., sense). In the early phase of language development in the child (Brown 1973), the phase of nonidentical replication of concepts can be easily recognized. The double system of conceptual thinking —linguistic expression—appears in the phase of quasi-identical replication.

It is well known that children's acquisition of a first language is fast and easy, while learning a second language is usually slow and probably occurs by a different learning mechanism.

This phenomenon can be interpreted within the autogenetic model. In learning a first language the main organizational role is played by the simple replicative cycles connected to the "self," that is, the structures corresponding to Piaget's second phase. The child immersed in the linguistic environment of adults acquires language structures, which are incorporated into its primitive concept-replication cycles, thus extending them. In other words, the child operates the environment by its use of language; something or somebody is carried to or removed from the self (the child) by the use of language structures. This mechanism draws the boundary conditions for the linguistic patterns that children are able to acquire. Further language learning extends replicative cycles of concepts with the third and further Piaget phases. Following the rapid acquisition of a first language, the brain contains a complex linguistic "concept ecosystem." The structure of a second language can join to the replicative cycles only through the existing and identically replicating linguistic construction, which inhibits and slows down learning of a second language. However, we may assume that there is an optimal learning process that could join the structures of a second language to the replicative cycles very quickly.

At present it is difficult to assess the accuracy of the replication of concepts, but it seems probable that it does not even approximate that of molecular systems. The various means developed by man—formal logic, mathematics, scientific methodology, etc.—all serve to increase the relative accuracy of concepts. The phenomenon of *compartmentalization* also can be recognized in human thinking. For example, in the course of personality development, separate domains may arise, the concept components of which are harmonized first of all within each domain. The quasi-identical replication of concepts leads to *convergence*. This feature of the autogenetic system implies that the replicative information converges toward a maximal value, and when this value is reached all parts of the system behave in coordination as a new replicative unit.

The next example is only a roughly outlined illustration of this phenomenon in human thinking. A relatively long period in the child's intellectual development is characterized by its tolerance of logical contradictions. The very essence of development is the gradual elimination of these early contradictions—the formation of some kind of worldview (Piaget 1954). In this process the different concepts become coordinated. Formally, this phenomenon is very similar to ecological succession. The coordinated concepts coreplicate, forming increasingly larger replicative units. A brain well advanced in convergence does not accept new ideas without criticism, and it rarely forms completely novel views if the new concepts cannot be integrated into the "ecological network" of its existing concept population. Adult, "mature" individuals can reach a very high degree of convergence where their capacity to absorb new knowledge is reduced. Their old habits, views, principles, or emotions can hardly be changed. A concept network has developed that is coordinated in all functional respects and behaves almost as an independent replicative unit. The assimilation of new concepts is only temporary and either of low efficiency or not at all possible.

The *organic concept* of memory was developed in the 1970s by the psychological school of Geneva, based mainly on experiments in which a subsequent "improvement" of children's memory could be demonstrated (Piaget and Inhelder 1973). According to Piaget, memory is capable of structural reorganization and is not simply "engraved" but is formed in close interaction with the developing intellect, undergoing a kind of ontogenetic, "organismic" developmental process. The experiments, as well as their explanations, were received with a certain skepticism (Liben 1976), mostly because their theoretical basis was unclear. The replicative memory model provides full explanation of these phenomena.

Parametric and replicative information in the autogenesis of the human brain

The parametric information content of the brain is practically 100 percent when concepts generated in the brain do not influence one another's probability of genesis. This is very likely the case at the end of embryonic development, immediately before birth. The parametric information consists of concepts developing as a result of primitive, environmental, stimulus-response connections, under the primary influence of species-specific inherited factors. The interaction of concepts appears very early in the growing child's brain. This is quite obvious in the development of secondary and tertiary reactions. Complex feedback relations develop between individual concept structures, and from this time on the brain's functional information content is continually growing. It is obvious that a great proportion of the information content of concepts in the adult brain is functional information, mainly technical, mathematical, etc., knowledge.

The duality of functional information predominates as well. In addition to the easily traceable special functions, the *replicative function* and, simultaneously, *replicative information* appear, enabling continuous replication of those components in which it is embodied within the given system.

In the course of convergence of the neural autogenetic system, replicative information is becoming organized in a hierarchical system in which concepts concerning the "self" occupy the highest regions. Replicative information grows constantly in the course of ontogenesis, reaching its maximal level through convergence when the whole system as a replicative unit is renewed in a coordinated, continuous, and unchanged manner.

7 Cultural Evolution

The idea of evolution at levels higher than those of biological organization was developed long ago in the social sciences. The first detailed theory of cultural evolution was drawn up by Spencer (1889), who recognized the continuation of biological evolution in "superorganic" organization.

Today "cultural evolution" is an accepted term of cultural anthropology (Alland 1973), although there is no generally accepted definition of culture. Chapple (1970) defined culture as "the totality of learned environment within which a particular group of human beings adapt." Generally, culture is considered to have three constituents: social, material, and mental (Osgood 1959). Social culture comprises the interrelationships of people; material culture is concerned with the production and use of artifacts; while mental culture involves those ideas not manifested in the other two. Developmental processes are always determined by both genetic and environmental factors. In the study of cultural evolution a strong structuralistic attitude prevails. There is an attempt to define the smallest cultural "units" that make up more complex phenomena. According to Taylor (1969), this is the concept of the "cultural trait," which is a socially accepted custom or activity that can be interpreted by itself without its environment. From these smaller units various cultural actions and complexes are formed in which the building blocks—i.e., cultural traits—are organically interconnected. Any kind of cultural structure, artifact, social connection, or idea can be traced back to the *material factors of the individual nervous systems.*

We will not discuss here those cultural theories that regard the emergence of culture as entirely independent of biological systems on the assumption that culture is a self-contained, independent system that cannot and must not be seen as a continuation of biological evolution. Other hypotheses, which are based on the assumption of a close and fundamental relationship between biological and cultural evolution, will be presented.

Animal cultures

There are many well-known observations of animals sharing behavior patterns through imitation (Bonner 1980). Some authors consider these phenomena to be animal culture. Very rarely can learning by imitation be observed even in insects—for instance, the dance language of bees. After returning to the hive bee scouts perform a well-defined moving pattern, the "dance," that in turn is imitated by some other bees; later the imitators themselves fly away to find the food source's location (Frisch 1967).

Hinde and Fischer (1952) have observed that titmice can learn from each other how to open the plastic caps of milk bottles left in front of houses. It also was noted that this new habit spread in concentric waves from its place of invention. In birds as well the "cultural inheritance" of certain song patterns can be demonstrated. A special pattern may outlast by generations the individual that first produced it (Jenkins 1977). In higher animals, especially in monkeys, similar phenomena have been found—first of all the cultural spreading of different "technical skills" (Kawamura 1963, Kawai 1965).

These kinds of behavior are called "traditions" in ethology. In order to accept behaviors as traditions, their spreading by imitative learning, their transfer between generations, and their variability within populations of the respective species must be demonstrated.

Mundinger (1980) has collected more than forty cases of animal traditions, a small fraction of those which actually may exist. However, he does not consider every tradition to be an expression of animal culture. His definition of culture is as follows: "culture is a set of populations that are replicated generation after generation by learning —an overt population of functionally related, shared, imitable patterns of behavior (and any material products produced) and, simultaneously, a covert population of acquired neural codes for those behaviors." This definition can be applied to human culture as well; it also is very close to our earlier, independent concept of culture (Csányi 1978), and it can be applied to some forms of animal traditions. In song learning by the house finch, for example, song patterns are transmitted from generation to generation by vocal imitation, and individual phrases form a functionally organized system. This organization is population-specific; in other words, there is a population of behavior units that fulfills the Mundinger criteria. It also has been shown that directly unobservable populations of neural codes exist in the brains of finches in the form of *memory traces* (Nottebohm 1977). Song learning of finches and other birds therefore can be seen as culture, while other, simpler cases of animal traditions, such as the dance of bees, cannot. It is evident (and also emphasized by Mundinger) that there is an enormous difference between the complexity of human and animal cultures. This difference, however, cannot be considered fundamental. Among the many kinds of animal culture, human culture is the most complex and the most important, but it is by no means unique. According to Lorenz (1973), cultural evolution is analogous with biological evolution, and it is amenable to the comparative evolutionary method. Cloak (1975) worked out principles of a "cultural" ethology, at the same time calling attention to the replicative nature of elementary interactions. There is no basic difference between animal and human culture, although human culture represents a new level of organization (as we will try to show below). In order to maintain this distinction, animal cultures will be called *protocultures*.

The question of the genetic determination of culture: coevolution of genes and culture

The concept of an evolutionary continuity between biological and cultural organization assumes that both the animal protoculture and the human culture result from definite genetic changes. There is (may be) a connection between cultural traits and the genetic architecture of the given species. This statement is a special formulation of a comprehensive view that assumes a connection between the social behavior of animals and man as well as a connection between their genes.

Sociobiology deals with these general questions (Hamilton 1964, Trivers 1971, 1985, Wilson 1975) and has provoked hostile arguments during the past decade (Sahlins 1976, Caplan 1978, Kings College Sociobiological Group 1982). Basic agreement exists concerning the evolution of the social behavior of animals (cooperation, group forming, etc.), and disputes mainly concern mechanisms. However, there is no general agreement regarding the emergence of human culture, and extreme views range from a culture that is genetically determined to a complete rejection of any biological influence on culture.

Sociobiology has been established by the analysis of phenomena difficult to explain within the framework of classical Darwinian theory. The evolutionary explanation of the behavior of social insects, for example, caused much trouble even for Darwin. Worker ants and bees give up reproduction entirely "in the interest" of the colony. If the unit of selection is the individual, as has been supposed by classical theory, then such a trait must be selected against and cannot survive during evolution. Modern theoreticians use the term altruistic for behavior in which an individual uses energy or risks danger and consequently reduces its own reproductive success while increasing that of a conspecific or a group of conspecifics. Alarm calls of various animals, for example, are a kind of altruistic behavior. Lionesses nurse all cubs belonging to the pride without distinction, and male macaques form coalitions, the members of which readily help one another in cases of danger. "Helpers" can be found in many species. They usually are young animals adopting their siblings or helping their parents to raise the next brood. The evolution of altruism was first satisfactorily explained by Hamilton (1964), using the concept of inclusive fitness. Several predictions of the theory were verified, especially in social insects (Trivers and Hare 1976). In addition to the concept of inclusive fitness, group selection was used as a principle for explaining the evolution of altruism (Wynne-Edwards 1971). According to the group selectionist theory, kinship is not a precondition for the emergence of altruistic behavior and can satisfactorily be explained by selection acting among *groups*. In reality the theories of inclusive fitness and group selection overlap and cannot be clearly distinguished. The pure form of group selection, in which kinship relations are ignored, is called "inter-demic selection" (deme is a reproductively more or less closed unit of the population), while the theory considering inclusive fitness connected with kinship is called "kin selection" (Maynard-Smith 1964, Williams 1966, Wilson 1973). The concept of inter-demic selection has often been attacked, and in reality it has proved difficult to find conditions fulfilling the rigid criteria that follow from the theory. Interestingly, the most convincing proofs were found in the case of man, where in the early stage of evolution the existence of small, closed groups and their competition and selective extinction can be regarded as the main forces of evolution (Alexander 1974, Alexander and Borgia 1978).

The theory of *reciprocal altruism* is connected again with individual selection (Trivers 1971). It explains the evolution of altruistic behavior by reciprocity among unrelated individuals that leads to the increased fitness of both participants. The concept of reciprocal altruism was first raised by Kropotkin (1902) and was revived by new theoretical considerations (Brown et al. 1982, Aoki 1983).

A scientifically sound area of the sociobiology debate concerns the level of selection (whether it is the individual or the group) and the origin of altruism. Genetic adaptation was taken for granted by many sociobiologists, and they tended to explain the existence of social traits by an "adaptationist story." Lewontin's (1979) sharp criticism showed the fallacies of this explanation. It cannot be supposed that all traits result from genetic adaptation. The debates were especially hot in the case of man, whose phenotypical units are rather difficult to define and on whom genetic experiments cannot be performed. In addition, several theoreticians who deny any role for biological factors in man's behavior directed the debate toward unscientific extremes.

In our opinion, most problems in sociobiology are caused by its proponents' ignorance about organizational levels; they disregard higher-level organization and attempt to explain phenomena by means of the dynamics of the lower hierarchical levels. As we have discussed, this kind of reductionism, while it can lead to the discovery of causal connections, is unsuitable for creating explanatory models.

The emergence of the coevolutionary theory is a way out from the crisis of sociobiology. Durham (1978) argues that we need not suppose a direct connection between cultural phenotypes and genes in the sense of a gene contributing to the formation of a given morphological phenotype. It is enough to assume that the emergence of cultural traits has been the consequence of the evolution of genes responsible for the *capacity of culture*. This means that culture is adaptive independently of its particular forms, and genetic selection is required only to ensure the evolution of physiological mechanisms, such as well-developed learning ability, large memory, social attraction, sophisticated communication, ability to imitate, etc., necessary to acquire culture. If these mechanisms exist, then some kind of culture will develop that will be an adaptive, organization-enhancing, reproductive success. This idea also would account for the great variability of cultures. The emerged culture, in turn, influences the group structure and reproductive mechanisms and feeds back to the genes; that is, a coevolutionary process begins. In this coevolutionary process, natural selection is replaced by cultural selection; that is, individuals best adapted to the cultural environment will have the greatest reproductive success.

A detailed treatment of the coevolutionary theory was published by one of sociobiology's founders and his colleague (Lumsden and Wilson 1981). They consider culture to be the totality of artifacts, behavior patterns, institutions, and mental constructions acquired by members of the society through *learning*. Phenotypic units of culture are named *culturgens*. According to the authors, a culturgen is not a genetic but a cultural concept, which basically corresponds to the cultural trait used by anthropologists or the memes of Dawkins mentioned earlier. (This is a rather uncertain definition because the cultural trait is a behavioral phenotype; the meme is a neural. Mundinger's definition including both "populations" seems more appropriate.)

In certain cases of ceremony or sexual custom—for example, the incest taboo—culturgens are discrete units and few in number. In other cases there are many culturgens, and the differences among them may be continuous—for example, the categories of color perception cited by Lumsden and Wilson. At any rate, the culturgen's most important property is its perceptible or conceivable unit character.

A further assumption of this culture model is that members of a cul-

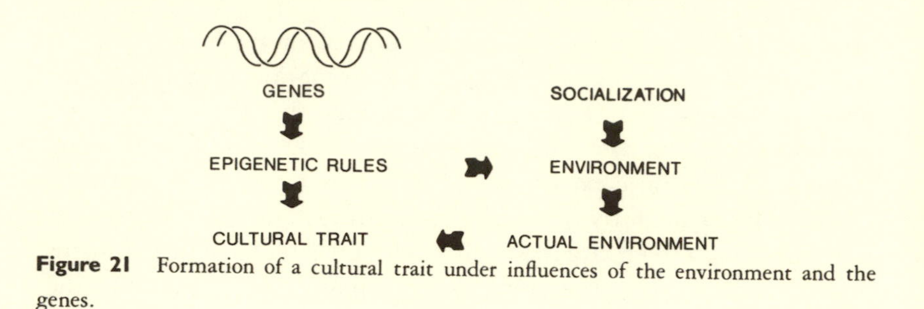

Figure 21 Formation of a cultural trait under influences of the environment and the genes.

ture acquire culturgens through *epigenetic rules* during socialization. The epigenetic rule is an old genetic concept that covers processes controlling the development of behavior and morphological traits during the organism's ontogenesis. Developmental processes always are determined by both genetic and environmental factors. The same genetic background may lead to different phenotypes in different environments as a result of interactions of genetic and environmental factors expressed in the epigenetic rules (figure 21).

Lumsden and Wilson argue that the probability of acquiring a given culturgen is determined by the epigenetic rules connected with behavior. Since culturgens are influenced by both environmental and genetic factors and their ratio may change, cultures determined solely by genetic or environmental factors are theoretically possible. However, it is more likely that genetic and environmental factors coevolve, and both contribute to the formation of culture. Epigenetic rules that are the most successful in a given culture, i.e., which control the socialization of individuals of the highest reproductive success, will predominate in the population. In this way a positive feedback mechanism emerges; in other words, genes influence the features of acquirable culturgens and vice versa. Thus both the genetic and cultural components of epigenetic rules are selected in each generation.

Interactions of cultural and genetic factors were formalized in various mathematical models by Wilson and Lumsden. Their most interesting results estimate the evolution rates of culturgens and the epigenetic rules (genes) behind them. It was calculated that relatively few generations (about fifty) allow the most successful epigenetic rules to spread through the entire human population. A firm establishment of this estimate would suggest the possibility of considerable genetic evolution in such a short time (e.g., the period of mankind's written history). Moreover, Lumsden and Wilson suppose that such a quick genetic change was actually a *precondition* for the development of modern cultures.

The above theory is based on several extremely strong assumptions, but it has the advantage of being falsifiable by appropriate means. To examine the validity of this theory, appropriate methods of recognizing and measuring epigenetic rules have to be worked out; also the connection between reproductive success and the variability of epigenetic rules must be established. Demonstrating a complete or even a partial genetic determination of culture would considerably change our present view of societal evolution's mechanisms. Verification or rejection of the theory is desirable, even if the existence of a strictly genetically determined culture seems unlikely.

Cultural complexes as second-order organisms or sociocultural machines

A very promising line of investigation of cultural evolution has been developed by Cavalli-Sforza and Feldman (1978, 1981) through the application of quantitative mathematical models used in classical genetics. In their opinion, the same basic phenomena of biological evolution can be found in cultural evolution. An innovation in culture, for example, corresponds to a *mutation* in genetics, and they regard it as a failure in copying or imitation. *Selection* may be stabilizing or neutral. The former can be illustrated by the developmental process that results in an optimal technique, and the latter is involved, for example, in the Catholic dogma of the Virgin Mary's Assumption. Developing a good technique can provide a considerable advantage, while belief in a particular dogma is neutral. Belief in a dogma belongs to the ideological framework of the church, and its spread depends on the church's influence; its acquisition in itself does not offer the individual any selective advantage. *Random genetic drift* is well known in population genetics; its equivalent in cultural evolution leads to the acquisition of rare techniques or beliefs by small, closed populations. *Transmission* analogous with the transfer of genetic material (DNA) in biology also can be found in culture in the form of copying and learning various cultural traits. Of course, while the analogies are obvious, the mechanisms may be entirely different.

In the case of transmission, ten modes are distinguished by Cavalli-Sforza and Feldman. There is cultural transmission from parent to child. For some traits the transmission may be uniparental (mode 1), while generally it is biparental (2). Other adult members of the population (3) contribute to the child's upbringing (servants and relatives in the parental generation). Remote generations contribute through written or oral tradition (4); this contribution will be made mostly by grandparents. Peers are sources of information (5), and another transmission mode is represented by the teacher-pupil relationship (6). There are two modes of political transmission. In small communities decisions are made, and the spread of information is directed by one or a few people (7). In a bigger political organization there is a social hierarchy characterized by the consensus of larger groups—for example, political parties—providing an efficient way of spreading information (8). An obvious mode of cultural transmission is realized through the mass media (9). Finally, all societies are organized in relatively isolated groups, which partially overlap or communicate. Between some groups, however, communication may be minimal, creating opportunities for independent cultural evolution (10). Quantitative models were applied experientally in analyzing various cultural traits such as beliefs, customs, and activities. It was found, for example, that religious or political beliefs are mostly transmitted within the family (Cavalli-Sforza et al. 1982).

A similar concept of the evolution of ideas (called abstract evolution) was worked out by Lynch (1983). He examined the relationship of ideas and their "host," identifying conditions in which either the idea-carrier hosts are propagating faster or the transmission of ideas is especially favorable. He made a penetrating analysis of the two-hundred-year history of the Amish, a closed religious community in the United States. (See an excellent monograph on the Amish by Hostetler, 1980.) The Amish community is absolutely closed. It originates from German immigrants exiled from Europe because of their religious beliefs. Their reproduction rate is very high because of various religious taboos. They do not use modern machinery such as telephones, radios, or cars, and one of their main convictions is that the Amish community must be isolated from its non-Amish neighbors. As a consequence, the Amish ideology remains intact since all reformers are excommunicated as heretics. Despite a loss rate of about 20 percent per generation to the attractions of contemporary American society, the Amish population (the carrier of Amish ideas) is increasing.

All abstract models of cultural evolution are based on the assumption that various cultural complexes are individual entities possessing a kind of stable organization, and their interactions can be compared to biological evolution. Cavalli-Sforza and Feldman (1981) have gone so far as to regard cultural complexes as *second-order organisms*. We find this concept very important with respect to our own model. Therefore, we want to make some further comments.

Mumford (1967) was perhaps the first proponent of the concept of living "machines," which are assemblies of human individuals created by appropriate training and discipline. These machines can produce various artifacts, or they may serve military, religious, or other purposes. The "machine" is principally a programmable device, a team of experts that can perform tasks according to plan, should it be production of something or an action of some purpose. A similar concept was developed by Ellul (1965), who considered a "technical complex" as a functional unity in which a set of inventions had been combined and which showed the properties of self-maintenance and self-development. Finally, Polányi (1968), who formulated a general definition of machines, must be quoted again. He considered a machine to be every kind of organization in which *constraints* of a higher level were built on the dynamics of the components of lower levels. In this regard there is no difference between mechanical machines constructed by man and biological organisms. We may add to this that the above definition also is valid for *cultural machines*.

In what follows we regard as a *cultural machine* every cultural or technical complex that has a material or mental product. If the product of the machine is *itself*, as was found in the case of organisms by Maturana and Varela (1980), then the machine is *autopoietic;* if the product is something else, it is *allopoietic*. For example, a cultural machine is a traditional university or school in which idea carriers are produced in a controlled and planned teaching process and the most standardized products are favored. Factories also are cultural machines, their managers and workers able to operate particular technologies and to produce given artifacts. Cultural machines are characterized as operating by idea schemes, working on preassigned tasks. Their "parts," humans, can be replaced—at least gradually. A new "part" can operate properly only after acquiring the idea network that characterizes the given machine. In cultural machines called institutions, processes of discipline can be found that provide stability for the idea network.

In a society individuals are always faced with cultural machines. Employees of an institution are more or less passive carriers of the machine's idea network. Cultural machines are analogous to living organisms in many respects, but they also can be considered supercycles within our autogenetic model, which give structure and organization and which are self-maintaining. They are *replicative component systems* that attain reproduction and self-maintenance by replicating their own components. The origin of cultural machines can be easily described by autogenesis.

The autogenetic model of cultural evolution

The autogenetic model makes it possible to examine the whole human culture as a single system. It provides an explanation for culture's origin, organizational levels, and future direction. Human culture has all the necessary characteristics for it to be regarded as a *component system*. Its components are complex structures, as there are several organizational levels below culture. Its components belong to three classes:

(1) *Living organisms*. These include all biological organisms that contribute to social interactions in a human society. More precisely, it means humans and organisms that supply food, raw materials, microorganisms, parasites, etc. Organisms that do not belong to culture—for example, plants producing oxygen—are not components of the cultural system but belong to the conditions of culture, being an essential part of its environment.

(2) *Artifacts*. All formed, processed objects that were changed by human intervention are regarded as artifacts. There is some difficulty in classifying artifacts and natural objects. A mountain and a pebble are natural objects in their original locations, but a pebble carried to

a museum and exhibited there is regarded as an artifact, although its alteration resulting from human intervention is small. Ores also are regarded as natural objects in their natural location, but after mining they become artifacts.

(3) *Ideas*. Mental representations. Those smallest, intelligently definable actions or thoughts, determined by physical factors of the nervous system, which can be communicated, copied, or formed as artifacts, or performed as social acts, are called ideas. The creation of complex social acts or thoughts or artifacts occurs by contribution from an idea population. (The concepts of neural autogenesis are subunits of ideas. On the level of cultural evolution the interaction of several concepts forms an idea. In contrast with an idea, a concept is not necessarily communicable or performed.) Ideas belong to three categories:

(a) Social ideas. Man lives and operates within a network of complex social interrelationships. Social behavior is molded by socialization, imitation, learning, and discipline. In the brain of each individual a characteristic idea population is formed, which regulates its social actions. Ideas determine not only personal relationships, social institutions, and attitudes toward these institutions but various value orientations as well. In a closed group the social structure is formed from idea populations possessed by the brains of that group's members. Therefore, the determination of social structure is populational and stochastic in character. The individual as a carrier of a part of the idea population that determines the social structure is a creator of this very structure, which exists as a *reality* and represents his environment in the same sense as was formulated by Berger and Luckmann (1967) in their theory of reality as a social construction.

(b) Material ideas. This category contains ideas that are revealed by creation of an artifact. Even the simplest objects of everyday life (e.g., a spoon) represent ideas. These ideas are expressed in the function, value, and mode of production of each artifact. It should be noted that during the creation of an artifact, the object itself becomes a carrier of structural information. The idea population of the maker's brain is represented in the artifact and can be considered the structural information of the object. This information transfer is not a one-way process; i.e., the information located in the object can occasionally be transferred back to the human brain. Hence, human brains and artifacts appear as two continuously communicating compartments of the idea domain. These interrelationships can be observed readily during various social rites in which the presence of artifacts contributes to the formation of the proper social behavior. The objects used in such rites help the participants to remember, perform, and transmit the special sequences of the given behavior. Images and artifacts exchange information and develop interwoven functions. Interactions of artifacts with members of other idea classes create *symbols*, representing pure ideas, which function independently of the particular objects carrying them.

(c) Mental ideas. Here belong the ideas that are represented neither in artifacts nor in social relations, which makes this category rather arbitrary. For example, a mental idea might be the concept of future, infinity, history, mathematical conceptions, etc. Mental ideas, replicated by spoken language, may spread and propagate very quickly. The mental idea of a good joke can reach millions of people within days. Printing and mass communication have made idea replication increasingly rapid. There are strong interconnections among mental ideas and other ideas. For example, ideas of a tale are considered mental ideas, but the tale can be written or printed, and by this process the ideas become artifacts. Moreover, the same tale may contain social references; therefore, it may influence social interactions, so its constituent ideas may fit into the category of social ideas.

Composition and decomposition of the cultural system's components

Cultural evolution's components are continuously composed and decomposed as a consequence of the energy flowing through the system. (The concept of energy is used here in its physical sense.) In the class of living components the mode of production is replication because components belonging here also are components of the biosphere; thus metabolism and energetic processes still bind them to the biosphere.

Cultural components are produced by the operation of the human brain. Concepts (the subunits of ideas) are replicative structures, and the ideas themselves are replicated. Imitation and other processes of learning also are replicative. Fidelity of idea replication may be anything between zero and one. Ideas replicate during human activities in two senses. First, they replicate from person to person within the population, which is called horizontal transfer by Cavalli-Sforza and Feldman (1981). Second, ideas are transferred from one generation to the next. This is done by the various educational institutions organized to provide an idea replication of high fidelity. Cavalli-Sforza and Feldman call this vertical transfer. Replication of ideas is accompanied by "mutation," "recombination," and selection because available space among the population of human brains is limited. In other words, ideas are capable of evolution. (Similar characteristics of artifacts will be discussed in detail in the chapter on technical evolution.)

Society as a complex replicative network: creative spaces

Society can be regarded as a complex component system. Its components are cultural complexes, i.e., cultural machines: institutions, religions, factories, etc. The creation of components involves the action and contribution of other components. Creation of a complex object or idea requires the contribution of a fairly large set of components. To create a particular object, some raw material is needed; to get raw material, other objects are needed, etc. Therefore, we use a concept that appropriately reflects the complexity of the creative process. We call this *creative space*.

The creative space is a model, an abstract space in which a representation of each component involved in the production of other components is given. So we speak of technical space, i.e., the creative space of artifacts, of cultural space connected with the creation of various political and social ideas, and of biological space that creates living beings. The concept of creative space helps to recognize the mutual interaction of cultural evolution's components.

It can be shown that creative spaces have *replicative organization;* i.e., information represented in the created components feeds back to the process of its own creation. It is a *replicative type of information* that is generally more or less distributed in the creative space. Cars, for example, are created in factories where the information for production is located on blueprints. Information transfer from blueprint to artifact is seemingly a one-directional process. But if we examine the entire creative space of artifacts, it appears that the already prepared copies of the car *influence* the car-producing process and the blueprints in many ways. The most common process of design is *copying*. The designer copies and sometimes recombines reliable parts of previous production cycles. Even if he invents something, it may have originated from the creative process of a different artifact. Again, obvious examples are the earliest automobiles, which resemble the mail coach in most of their details, because designers copied the mail coach. After a while design of the chassis became increasingly different from the mail coach as the result of a slow evolutionary process in which imperfection of the copying process and small inventions accumulated.

The same process can be observed in the creation of various ideas. The shaped, written, or composed idea is often regarded as an exclusive product of the creative mind. If, however, we consider the entire creative cultural space, we can identify a number of already existing ideas that have provided the necessary information for the creative mind. Ideas are copied, recombined, and only to a lesser extent invented.

A further important characteristic of creative space is that various kinds of components are found together, which means that every component of the cultural system is capable of influencing the probability of genesis of others. For example, artifacts influence the probability of genesis of organisms in agriculture; the operation of mental ideas creates artifacts of fine art; social ideas influence the technological processes of industry; etc. Thus creative spaces form a *functionally and organizationally* united system.

Precursors of the cultural system

It has been shown that one precondition of autogenesis is the emergence of an autogenetic system precursor, which organizes the construction of special constraints on the dynamics of the organizational level already in existence. It also was shown that animal protocultures spontaneously and frequently appear at the level of biological organization. In this section we will examine the formation of human protocultures and put forward our assumption that these are the system precursors for cultural evolution. Hominid species living in complex social groups, using tools, and developing communication that helped them satisfy their biological needs appeared at an early stage in human evolution. Interactions among social relationships, tool using, and communication (or more precisely the *organization* of these interactions) formed the system precursor of cultural evolution —even though these interactions in themselves cannot be regarded as a cultural organization. The true cultural system was created by the continuous extension of the ancient system precursor by *autogenesis*.

The zero-system of cultural evolution and the biological basis of human culture

Evolution of human culture has been built on the dynamics of organization at the biological level. The early period of human evolution can be regarded as the zero-system of cultural evolution. A great many paleontological data recently have been collected that show major branching in the evolution of higher primates occurred 6 to 7 million years ago. The common ancestors of humans and chimpanzees lived around that time. Common ancestors of humans and gorillas are about 8 to 10 million years old, while humans separated from orangutans 13 to 16 million years ago and from baboons about 27 to 33 million years ago (Sibley and Ahlquist 1984). Since separating, a calculated difference of about 1 percent in nucleotide substitutions appeared in the genomes of humans and chimpanzees (King and Wilson 1975). It is not easy to decide whether this percent is a lot or a little, and accordingly opinions vary. There are differences in chromosomal structure (nine inversions) between the two species (Yunis et al. 1980). It is well known that a precondition for chromosomal inversions is close inbreeding, which indicates that evolution has been based on small populations since the first separation. The oldest fossils of Hominidae are 4 to 5 million years old and polyphyletic in character, although the expert opinion differs regarding their exact taxonomical identification. There is general agreement that *Australopithecus afarensis* probably was the common ancestor of the hominids and that later Australopithecine species also were bipedal, tool-using species hunting for large prey (Falk 1983). In the Homo line of descent, the tool-using *Homo habilis* is 2 million years old and already had a large (600 ml) cranium. *Homo erectus* appeared about 1.6 million years ago; its cranium was about 900 ml and besides tools he also used fire (Gowlet et al. 1981). *Homo sapiens* is about 600,000 years old with a 1,400 ml cranium (Cronin et al. 1981).

Revelation of the biological basis of cultural evolution can be expected mainly from examining characteristic differences among humans and other higher primates (Isaac 1978).

The use of language

Social animals have well-developed communication systems. (Certainly a complex group structure cannot develop without communication.) Units of animal communication are definite, specific patterns of behavior, which usually designate internal states or intentions or, in special cases, convey information about the environment (Wilson 1975). Animal communication is analog communication and is closed. A behavior pattern is a communicative signal if it influences the behavior of the recipient animal in a way that is adaptive to the sender.

Highly developed human sociability is accompanied by the development of a unique means of communication—human language—that has no parallel in animals (Hockett 1960). There are four main features of human language: (a) It possesses a dictionary, interpreted uniformly by those speaking the same language. (b) The signals of the dictionary are symbolic; that is, they have no connection with the physiological or behavioral processes related to the word's meaning. An animal can express aggression only by simultaneously being aggressive, so its signals are not symbolic. (c) The dictionary's symbolic signals can be expressed in combinations of different meanings by using definite rules. The number of possible combinations is theoretically infinite (Glaserfeld 1976). That is, human language is open. Finally, a most important characteristic of man in connection with language is his *capability of displacement* (Brown 1973). (d) This term implies that man is able to reduce a situation perceived through his senses to its components—to analyze it—and in the course of analysis to create a new structure from the components. In other words, he is capable of a kind of linguistic-logical synthesis. With the help of "reconstruction," man can displace phenomena in space and time, change relationships, create *linguistic models of reality,* then operate and analyze them as well. The "capability of reconstruction" is obviously not identical with language, but it is manifested through the use of language, being the most specific feature of human thinking. Displacement is made possible by the identical replication of concept components. The concepts evolving in the phase of nonidentical replication are soon deformed and merge. The separation of a single phenomenon of the environment and its association with another phenomenon, or a group of phenomena, at a much later time or in an entirely different environment becomes possible only through identically reproduced memory structures.

The most important constituent of cultural evolution is an individual's ability to transfer the concept components of his own memory (by copying it with varying fidelity) into the memory space of another individual through language. This copying process is *replication,* that is, propagation of the concept components in a physical sense. As a consequence, the evolution of concepts is no longer restricted by an individual's life span. Individuals living in a linguistic community join their memory space and create a replicative space much larger than individual ones. With this new mode of concept replication, an autogenetic process begins. *Parallel connection* of the modeling brains makes possible the creation of new types of environmental models. These can be called *supermodels.* The supermodel is not based on the sole experience of an individual but is the product of a continuously communicating group. It was found that even in the most primitive societies the time necessary for getting enough food and satisfying basic needs is no longer than twenty-five to forty hours per week. The remaining time is spent in various cultural activities, mostly *talking* (Lee 1969). Conversations serve to create and refine the supermodel of the environment within the group. The knowledge necessary for individual and group survival, the location of various resources, the way to get and use them, social structure and identity of the group, myths and religion, all are parts of the common supermodel. New generations receive the supermodel, embodying the experiences of many generations and past events. The supermodel is a dynamic structure, a system in itself, which appears as *social reality* for individuals. As a consequence, the supermodel is separated from its carriers (individuals) and operates at a new organizational level.

We should paraphrase Pask's theory of talking (1975). He essentially states that *understanding* emerges in the course of talking if a model develops in the brain of each partner, each of whom is capable of predictions similar to the other's concerning the subject of their talk.

Man's attraction toward objects

Several animals are known for their particular attraction toward certain objects in their environment. Good examples are the nest-building of birds and the construction of dams by beavers. Animals use objects not only as raw materials but as tools. Vultures or mongooses use stones to break up eggs (Goodall and van Lawick 1970); the sea otter uses stones to open shells (Houk and Geibel 1974); baboons and chimpanzees apply sticks, branches, and stones for more than ten different purposes (Goodall et al. 1973, MacGrew and Tutin 1973, Teleki 1974). Although the number of reports about animals using tools is always growing, the use of tools and the attraction to objects can be regarded as rare phenomena in the animal kingdom. In contrast, no human culture is known, however primitive, that could not be characterized by the permanent use of at least a few self-made tools (Taylor 1969). Humans have an affection for objects; man being active without objects is a rare exception. Undoubtedly, this attraction is based on genetic programs, as it can be observed in very early childhood. Aside from social contacts, the most important activity of the developing child is the examination of objects within its environment (Piaget 1952). According to primatologists, this characteristic is an early evolutionary inheritance of man, originating from the fact that his remote ancestors still lived among the *branches of trees* and developed as a result of the adaptation necessary for locomotion among branches. When moving among branches, primates make contact with "objects," that is, branch segments of different shapes, sizes, and textures. Our primate ancestor, "descending" to the ground during the evolutionary process, had inherently in its "hand" (or rather in its brain) the intelligent strategies of handling objects and most likely its special attraction toward objects, which is unmatched in its dimensions among other species (Jerison 1973).

Carrying objects also is characteristic of man. An artifact is not thrown away after its making and use but is carried by its maker or user. Chimpanzees look for a useful object in the very place of the intended use, and they throw it away after they use it. This characteristic is probably connected with the development of another property, that is, carrying the prey home and not consuming it immediately after the hunt.

Social structure

The organization of mammals is usually characterized by mating types, ways of rearing progeny, methods of foraging, and structures of groups (Eisenberg 1981). Most mammals are polygamous; obligate monogamy is rare, accounting for about 10 percent of all species. The progeny is usually reared by the female, and male participation is low. Methods of foraging are enormously variable, the dominant type being the solitary hunter and gatherer (82 percent). There also is a great variability in group structure, ranging from pair-bonding restricted to the short mating period through the nuclear and extended family to complex troops (Eisenberg 1981). In the social behavior of our closest relative, the chimpanzee, many characteristics reminiscent of human behavior are found. The members of a troop form smaller subgroups, usually consisting of males, or a mother and her young. Males are the residents (that is, young males remain with the group), but females leave the territory. Therefore, kinship among males is usually closer. Temporary male subgroups vigorously patrol and defend the territory. Upon meeting males of alien troops, they become very aggressive; murder of strange males is frequent. Males also are aggressive toward strange females, but they never take their lives; on the contrary, they try to prevent their leaving the troop's territory. A dominance order exists among the males, but aggression within the group is not very strong. Males have a high tolerance for each other; moreover, male coalitions and mutual help are common. Sexual competition is strong, especially in small groups, but exclusive primacy of a single male is rarely found in larger groups. In free-living chimpanzee groups, a male consort of the female in oestrus is frequent; the pair tends to range apart from others for a couple of days, forming a kind of pair-bonding. The young are thoughtfully cared for by females over a long period. Males do not participate in the care of the young, but they show apparent tolerance toward them (Symons

1979, Hinde 1983). Group hunting for small animals, usually monkeys, is common. In the hunt, only cooperating males participate, but sharing the prey does not exist except for "tolerated scrounging." The individual actually catching the prey tolerates others taking smaller pieces of his food (this is the way females get a share), but as a rule he tries to withdraw with his prey and eat it alone (Teleki 1975).

An intensive study is taking place aimed at reconstructing the biological features of human evolution from paleontological (Isaac 1978) and comparative evolutionary data (Passingham 1982, Hinde 1983). Although there is no general agreement as yet, the main features seem to be established. Polygamous and monogamous marriage systems occur among various modern societies in a ratio of 3:1 (Murdock 1967). The biological basis for polygamous tendencies is supported by the sexual dimorphism found in man. It seems that the present situation is characterized by great genetic and cultural variability. This statement also is supported (at least concerning sexual behavior) by the well-known facts that in monogamous societies polygamous tendencies survive in various forms, such as prostitution, open marriage, concubinage, etc. In polygamous societies deviations toward monogamy also exist in the forms of primary wives or other kinds of privileged position among legal wives. Since mammals, even primates, are predominantly polygamous (with the rare exception of gibbons, a branch of apes farthest from man), it is very likely that in the early period of human evolution polygamy was the dominant form. But soon conditions appeared that, to a certain extent, favored monogamy. It frequently is pointed out that 90 percent of birds are monogamous because of the high cost of parental care. Egg-laying, nest defense, and feeding the young require the participation of both parents (Passingham 1982). In human families of either polygamous or monogamous societies, fathers always have important roles in raising children. Therefore, it is probable that a shift from polygyny to monogamy resulted from the need for long care of the offspring and cooperation within the protohuman groups.

Among chimpanzees, mating between mother and sons is very rare. In human groups there is a pronounced incest taboo and organized exogamy (Murray 1980). A plausible explanation for these phenomena is that early human evolution has been going on in small groups, and exogamy developed as a cultural mechanism to avoid inbreeding.

Sexual behavior in animals serves solely for reproduction. Man is the only species, except perhaps for dolphins, in which rejoicing and pair-bonding sex roles have evolved (Mellen 1981). Man is characterized by strong pair-bonding, but attraction exists not only between members of the opposite sexes. Man is exceptional even among the higher apes for the extreme intensity of his social affections. Man is willing to sacrifice whatever is necessary, even at the risk of his life, for group mates. Social attraction is especially well expressed in early childhood. Children are willing *to share their food* (Eibl-Eibesfeldt 1979), something that never occurs among even highly social apes. Care of offspring is characterized by its length and thoughtfulness as well as by the participation of group members other than parents. Mechanisms of social learning are especially well developed.

Man is the only species that recognizes and maintains various *kinship relations;* this is characteristic of all cultures. The basis of kinship may be genetic or cultural descent, but a definite and organized system embodying most of the group's interactions is certainly characteristic. Not even the most developed animals have such a kinship system. Among apes, for example, mother-young or at most sibling relations are recognized.

Precultural man was characterized by strong *cooperation* expressed in common hunting. Common hunting involving cooperation and understanding among the participants can be observed only in a relatively few species. It is most developed in canids, especially in wolves. Group hunting requires a highly developed brain because not only the prey's behavior but at the same time the movements and intentions of several group members must be followed and predicted. Concerted action may be taken only if an action plan exists or the events can be predicted with high precision. The leader must make quick decisions and evaluate alternatives, giving attention not only to its own but its companions' actions and purposes.

Precultural man carried the captured prey to his permanent home base and shared it with females, the young, and the elderly. Such sharing of food, aside from tolerated scrounging, and also the existence of permanent home bases where females, the young, and the sick remain and where common feeding occurs are not known among apes.

Characteristics of human protocultures and the precursors of cultural evolution

Emergence of cultural man is usually attributed to a single characteristic—for example, language, tool-using, or common hunting. Recent comparative evolutionary studies have shown that simplifying explanations are inadequate because all these traits depend on one another. On the one hand, they express the same ability; on the other, they are the components of a functionally organized, adaptive *coevolutionary process* in which they promote each other's development (figure 22).

It is increasingly apparent that the extremely high intelligence of man and apes has developed primarily in interactions within the *social environment;* other factors have played only a secondary role (Humphrey 1976). The habitat of protocultural man comprised the *group* as well. His survival and reproduction entirely depended on his connections to the group. The group's structure can be characterized by a mild male dominance. Members of the group cooperated in gathering, hunting, caring for their offspring, and defending the group. They were willing to share food and goods. Polygamy and sexual competition characteristic of the early period of evolution were mitigated by these other social traits. *Permanent pair-bonding* and monogamy consequently emerged.

A highly developed brain is a precondition for living in a complex social space. Evolution of the human brain's *displacement ability*

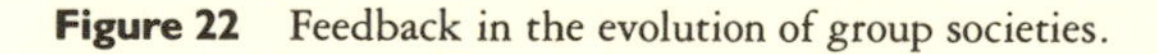

Figure 22 Feedback in the evolution of group societies.

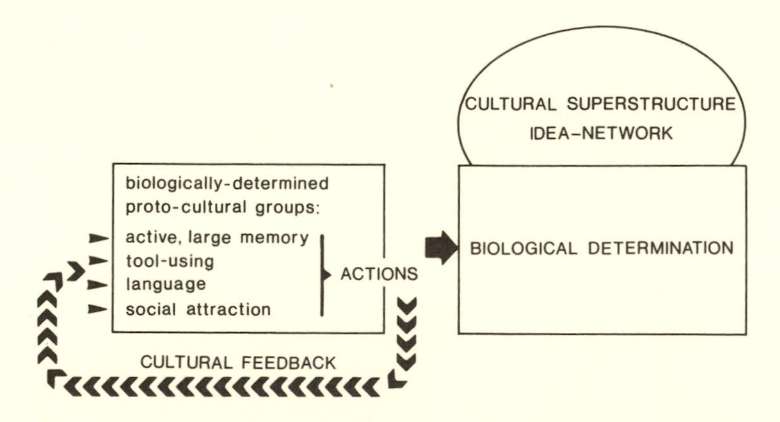

makes thinking in social relations and forming of a kinship system possible—the latter being the basis of the social space. Linguistic communication and tool-using also are expressions of displacement ability.

In recent years an anthropological hypothesis that supposes a definite and direct correspondence between tool-making, tool-using, and man's use of language has had a great impact on the study of human evolution (Hewes 1971). The notion of "creative intelligence" manifesting itself in toolmaking was already raised by Bergson (1902), but the considerable isomorphism of grammatical rules and toolmaking was recognized only in recent years (Montagu 1976).

The evolution of language produced communicable *ideas*, which later could raise the protocultural system to a higher level of organization. It seems clear that the formation of ideas, should they be concerned with the social, linguistic, or material sphere, have a *reinforcing* effect on the nervous system. Reinforcement is a concept of the physiology of learning; animals tend to repeat actions that have been reinforced immediately after they occurred. Reinforcement may take the form of food, water, cancellation of punishment, etc. This has long been known in connection with individual habits and stereotypes (Lorenz 1973). Repetition and permanence in social relations also reinforce existing connections (Tiger 1969). The same thing should happen in the development and repetition of linguistic and material ideas. The repeated, unchanged verbal recollection of a tale is indeed a pleasant thing. The creation of accurate idea structures for objects is facilitated and rewarded by the viewing, manipulating, and shaping of objects. For humans, therefore, the artifact is a means of creating and maintaining identical ideas.

The reinforcing effect of object manipulation and the creation of artifacts on the central nervous system already had appeared before the appearance of the human race. It was unambiguously demonstrated by Morris (1962) in the object manipulation of apes. Man strives to objectify various nonobjective ideas (fine arts, religion), which can be the basis of the creation of symbolic objects. It also can explain an otherwise striking phenomenon—the sudden increase in the variability of artifacts about 40,000 to 50,000 years ago, which were predominantly objects of symbolic content, i.e., jewels and cultic objects. According to Clark (1970), hardly any object used by man lacked a certain symbolic meaning. The common function of symbolic objects was to control and to guide social behavior into proper channels. In the evolution's early phase this function was manifested through the presence of objects as *primary stimuli* that help to preserve and pass on an idea having a social context.

In summary, the protocultural group was formed on the biological basis of sociability and the displacement ability of protocultural man. To satisfy basic needs, cultural complexes (including social and mental ideas) and objects were necessary. Social ideas shaped the group's organization; mental ideas included useful knowledge and myths; and artifacts were made for cultic or practical purposes. These *components* made survival and *replication* of the protocultural group in time and space possible. Therefore, *protocultural organization was the system precursor* of cultural evolution—the organizer of the autogenesis of human culture. Protocultural organization was formed by *replicative selection*, which established the interconnections among social relations, ideas, and artifacts, and its creative nature permitted the gradual extension of the primary organization as well as its high variability. Factors also emerged that, along with replicative selection, favored *competitive selection*. These will be discussed below.

The most characteristic feature of protocultural organization is the emergence of a special positive *feedback* that played an important role in human evolution's next phase. Cooperative hominids possessing relatively large brains and already using tools dominated their contemporary environment since no predators really threatened their groups. However, no evolutionary equilibrium or stasis was established (Cronin et al. 1981). The question remains: Why was that so? Which were the environmental or internal factors that induced the further rapid evolution that produced modern man with his linguistic competence, highly developed artifact and tool-making/tool-using ability, his symbolic thinking, cooperation and loyalty to his group for life and death, i.e., *Homo sapiens*?

There are many indicators of a transformation of the mechanisms of selection that resulted from protocultural organization. In the evolution of animals individual selection is regarded as most important, and the evidence for group selection is weak (Alexander 1974). Migration among animal groups is so high that it blurs the effects of the differential reproduction of groups, and so the unit of selection remains the individual. In the case of man, protocultural organization created the very conditions necessary for effective group selection (Eibl-Eibesfeldt 1982).

Language was certainly the decisive factor. Evolution of language and customs acted as *isolation factors*, having effectively mitigated migration among groups. Migration of a prelingual ape-man is constrained only to the ranging limits of contemporary apes. Troops are not very friendly to outsiders, but occasional joining is not prevented, especially in the case of females. As language and with it the other components of culture appear, it becomes more and more difficult for an individual to be accepted by an alien group. Migration was restricted enough to permit group selection. As a consequence, *antogonisms between groups* emerged. Linguistic isolation also created group consciousness, and with it a new social entity emerged, based not only on biological but cultural features. The groups of greatest cohesion that were able to defend themselves against others conquered the best territories.

Many data suggest antagonism among protocultural groups. It has been mentioned that the prey of the occasionally hunting chimpanzees are young baboons or other small monkeys. Chimpanzees did not become good hunters because meat is not fundamental to their diet, although according to Teleki (1973) it is rather important. Hunting activities of baboons also were observed. Obviously, early apes would have had a tendency to develop predatory habits if conditions had been appropriate. It also is clear that the easy prey would have been primate relatives because they were living in the same habitat and, being relatively weak, would have had no appropriate defense against the ape hunter. Quite possibly in the early period of evolution the protocultural hominids were specialized for hunting their fellow apes, and this could have been the very reason for the extinction of several species evolving at a slower rate (Roper 1969). The decreasing abundance of easy prey compelled the more developed species to hunt for other large animals, which increased the necessity for developing further cooperation. The emergence of language as an isolating mechanism also created a (group) selective pressure in the form of *active predation between groups*. As a direct consequence, the rate of evolution increased rapidly, and the properties of modern man developed.

The phases of cultural replication

The most important component of the zero-system of cultural evolution, protocultural man, is represented by the Homo line, which had been separated from apes. These creatures were living in small groups similar to those of the contemporary chimpanzee. Their developed sociability, verbal communication, and tool-using were the most important features of species-specific *protocultural organization*, which, as a system precursor, initiated cultural evolution. Perhaps it is unnecessary to show in detail that protocultural organization was replicative. The Homo species evolved in a stable environment for millions of years, and organismic reproduction ensured the reproduction of their groups as well.

Emergence of culture was allowed by the appearance of idea components, which could be copied and communicated from one generation to the next. In the following discussion we will look at the appearance of the various idea types.

Ideas of social relations

At the beginning the biological factors of social relations dominated these ideas. The social functions and roles developing within a group of people are closely related to the conditions of biological existence. As the replicative accuracy of ideas concerning social relations increases, the ideas can become more and more detached from their direct biological factors. In an exclusively noncultural group of animals the function of a dominant "superior" is largely determined biologically and depends on the individual's condition, experience, and previous conflicts. The culturally developed idea of a "superior," however, is interpreted only in terms of cultural traits. The person embodying it will not necessarily acquire this position by virtue of his biological qualities. The "rights" of a biologically determined superior—his potential and scope of action—are determined by biological relationships, while the scope of action of a culturally determined superior is governed by traditions, customs, ideas, i.e., the *culture* of the group. Therefore, the phenomena of the social sphere are always doubly determined: their basis and limits are prescribed by biological factors, whereas their contents, manifestations, and interactions are directed by cultural factors—*ideas*.

In a culture social roles become *institutionalized* (Hinde 1983). For example, the role of a chieftain or a king becomes independent of the individuals fulfilling this role over time. The cultural function of *power* joins the biologically determined dominance order and supresses the role of biological factors.

Although no direct data are available on the early evolution of the ideas of social relations, the initial state can be fairly well assessed by comparative evolutionary studies—first of all, by studies of the social relations among apes. The development of a culturally stable idea is the function of the accuracy of replication. For example, if within a group of chimpanzees some kind of social structure develops, this already will have numerous learned components, and the individual need not fight daily for its place in the hierarchy. As chimpanzees learn fast and remember for a long time, they retain their established positions far beyond their biological "limit of validity" (Goodall and van Lawick 1970). A change of position is not solely the result of individual actions of interested males, but that of the interaction of coalitions comprising several troop members (De Waal 1983). The established position would result in a culturally stable "idea" if there had been a means of preserving the learned materials for many generations, that is, if the physiological and mental apparatus of a quasi-identical replication of the idea were available. This apparatus is missing in chimpanzees.

In the course of his development man has become capable of the increasingly accurate replication of ideas. Initially, a social relation lasted only during the lifetimes of the individuals concerned, and new individuals created new relations. It is easily conceivable that the rapid development of social relations has adaptive value for the group, and that its unchanged continuation may provide a selective advantage, promoting the trend toward a more and more accurate cultural replication. Of the idea structures of extremely varied relations, only those will survive that have adaptive value in the given environment. Culture is not merely equal to the sum total of various customs, as has been stressed by Malinowski (1944). Individual cultural ideas have a function, satisfy definite needs, and interact with one another. This means that the idea structures can influence one another's probability of genesis, which opens the road to the whole system's evolution. The rapid acquisition of various social ideas promoted the group's survival. Therefore, genotypes producing a nervous system reconcilable with a *readiness to obey authority* gained adaptive value during the coevolutionary process (Milgram 1963, 1965, Eibl-Eibesfeldt 1982). This trait is completely unknown in animal behavior, but it plays an important role in every field of culture. It can be observed in early childhood that the child during its socialization easily and spontaneously obeys the rules of play and social life. The almost unconditional readiness to obey authority makes the organization of a complex culture possible, with all of its advantages and disadvantages. Rules are social ideas. In the propagation of ideas the *autocatalytic kinetics* (an essential phenomenon of evolution) characteristic of the structures at each evolutionary level can be clearly demonstrated (Jantsch and Waddington 1976).

As has been shown in the discussion of previous evolutionary levels, the appearance of functions of varying structures always leads to the emergence of larger complexes—the supercycles. Ruth Benedict (1934) in her famous studies on American Indian cultures has shown that in individual communities cultural complexes may develop that are completely different in each community but are in themselves coherent and coordinated. In those cultural complexes the various phenomena such as aggressiveness or mutual affection can be interpreted only in the context of the whole culture. As the process of cultural evolution is based on the replication of the concept superstructures—that is, the ideas in the brains of individuals—the general laws controlling their changes are no different from those operating at lower levels.

Ideas of language and the production of artifacts

The use of language itself is a certain kind of cultural idea system. Thus its development is bound to show the features of a real evolutionary process. The same applies to the idea components concerning the creation and use of artifacts. Isaac (1976) emphasizes two important components of the early evolutionary phase of various toolmaking activities. One is *differentiation,* the measure by which different shapes are separated in the mind of the toolmaker. The other is a system of rules, serving as the basis for producing the ordered assembly of forms, i.e., the artifacts. The development of both is made possible by a mental template, a concept evolving in the course of neural evolution. The first man-made artifacts were created in Africa about 2.5 to 3 million years ago. They show hardly any signs of design. For almost a million years these objects did not change; their shape was determined by the features of the raw materials rather than by the maker's plan. About 1.5 million years ago *symmetry*, the first unquestionably cultural idea, appeared (Isaac 1976). The first symbolic objects not serving direct practical purposes appeared only 50,000 to 100,000 years ago (Vèrtes 1964, Marshak 1976, Schwartz and Skoflek 1982). The development of language must have occurred between these dates, i.e., over a period of 1.5 million years if we accept that the development of artifact production adequately reflects the development of linguistic ideas. Considering either language itself or the development of artifacts, it is obvious that the appearance of individual idea components is immediately accompanied by the appearance of certain functions. The use of individual objects cannot be separated; they are interrelated. They act upon one another by their functions and influence one another's probability of genesis.

The same applies to linguistic structures. The individual units of language have special grammatical functions. These functions influence the probability of genesis of structures in the course of linguistic evolution. Development of certain grammatical rules "canalizes" the language, creating a definite linguistic superstructure. According to Bloom (1976), the phylogenesis of language is to a certain extent repeated in the period of acquiring language in childhood. The child is first preoccupied with objects in its environment, with its relation to these objects, and with its own social relations, and its first linguistic manifestations also are related to these things. It is reasonable to suppose that similar phenomena have occurred in the course of the evolution of linguistic idea components. The linguistic forms that developed among individual groups of people were initially loose, less specific, and with inaccurate replication and great variability. As the functional intertwining grew, there was an ever-greater need for exact differentiation, and this could be achieved only through increasing accuracy of replication. Therefore, preconditions formed by the cultural environment became favorable for the appropriate biological (genetic) changes of the nervous system, which serve a more precise idea replication. Fixed genetic changes made possible the formation of newer, more complex linguistic components, that is, grammatical rules. This is the very essence of the interaction of culture and biological organization, that is, *coevolution*, and this process has played a principal role in the development of language.

The origin of mental ideas

For the production of mental ideas a certain kind of ability—abstract thinking—is necessary, and from experiments performed on chimpanzees we know that they already have this ability (Premack 1980). Adequate use of the concept of "middleness" can be shown not only in chimpanzees but in lower primates (Rohles and Devine 1966, 1967). Rhesus monkeys, after an appropriate training process, are able to categorize pictures of flowers and insects they have never known (Lehr 1967). Thus we have every reason to suppose that the formation of mental and other ideas began simultaneously. Emergence of the idea of the *group* may have a great advantage. Although originally this is a social idea, in use it can be supplemented with the mental ideas of "interest," "survival," "origin," "life and death," "sacrifice," etc. These can form a very effective *functional organization*, which certainly enhances the survival of the group. Thus the coevolutionary process will promote the refined production of further mental ideas. Modern man is characterized by his ability for abstraction at a high level. For example, common mathematics belongs to the class of abstract mental ideas. It certainly would be naive to believe that mathematical ability in itself provided any advantage for humans during their evolution and that its emergence was adaptive. It is more likely that the ability to develop any kind of mental idea emerged because the abstract idea of the "group" represented a great adaptive value, and it served as an organizational center for shaping other mental ideas. Mathematical ability is only a by-product of forming abstract ideas.

As the various types of ideas emerged, a whole system capable of continuously forming ideas developed, which as a consequence of its replicative nature had the characteristics of an *autogenetic system*. Individual ideas can influence the probability of one another's genesis, i.e., *function*, and functional information appears. We can interpret both the parametric and functional information content of various ideas, but their more rigorous definition and quantitative measurement has not yet been accomplished. The parametric information content may be regarded as maximal in the period of the zero-system's development, that is, in precultural organization. At that time the developing components of cultural evolution have little influence on the probability of genesis of one another. The functional information content increases parallel with the interaction of ideas. For example, in the social sphere the social structures in which biological factors (e.g., the mother-child relationship) play the dominant role have high parametric information content, while in those insignificantly affected by biological factors (e.g., the structures of school systems) the parametric information content is small. The greater the differences between the complexes within individual cultures, the higher the functional information content of the complex components. As a consequence of idea interactions, this functional information is *replicative information*. In the idea components of cultural evolution the proportion of replicative information also increases.

In the early phase of cultural evolution replication of the idea components is of low fidelity because the neural mechanisms ensuring exact linguistic replication of the various ideas are missing. These mechanisms can evolve slowly through genetic changes during the coevolutionary process. Besides linguistic competence, based on genetical factors, cultural structures themselves also are able to influence the fidelity of replication. In the primitive culture of the Brazilian Auka Indians, who use only a few artifacts, Bauman and Patzelt (1977) found no generally accepted views within the tribe about complex abstract matters such as death, the existence of spirits, the origin of the tribe, etc. When such matters arose, everybody present gave his opinion; when *all* explanations, usually quite varied, had been presented, the given question was considered answered. Thus no well-established views were imposed on members of the tribe and transferred unchanged from generation to generation. This mechanism still reflects the nonidentical phase of idea replication, which later becomes more and more exact.

Processes characteristic of autogenesis can be clearly recognized. *Idea compartments* form—for example, ideas for making tools, special knowledge, "master skills," ideas of special hunting customs, myths, etc. Compartmentalization is accompanied by the functional organization of the ideas within each compartment, and as a consequence the fidelity of replication increases; that is, *convergence* begins.

In cultural evolution several phases of development toward identical replication can be recognized. At the beginning fidelity of replication is influenced only by the properties of the nervous system —mainly linguistic competence and ability to construct ideas by using language. A precondition for the nearly identical replication of ideas is that the human population of a given culture should not be too large so that ideas can spread by verbal communication without any constraints of time or space, and all members of the population should be able to participate in the formation of the common culture. If this condition is not fulfilled, the resulting culture will be separated into isolated compartments, which may be the basis of developing new types of culture. Thus the result of identical idea replication by verbal communication is nothing other than a cultural group, which will be regarded from now on as *group society*.

The exact timetable for the appearance of these societies is not known because protocultural hominids also have lived in small, *biologically* organized groups that are not yet cultural entities and where their existence has been determined exclusively by biological conditions. The autogenesis of ideas in these protocultural groups initiated cultural evolution and resulted in *group societies*. Group societies are replicative units not only in their biological aspects but in their cultural aspects. Ideas of the group—its artifacts and social connections —replicate from generation to generation within the limits of fidelity provided by linguistic communication.

The concept of group society satisfies the condition for being an *autopoietic machine*. Its components represent a functional network, which reproduces the very same network during its operation (Luhman 1982). The group society, as an autopoietic machine, has no inputs and outputs, creating only the conditions for its own existence, and it cannot produce a surplus to be traded, exchanged, or unevenly distributed; it is organizationally closed, stable, and in equilibrium with its environment. Emergence of this organization took several million years during human evolution.

The group society is an organization controlled by cultural constraints, built on the dynamics of the biological level of human existence. In its evolution replicative and competitive selection have played important roles. Replicative selection also is creative in the autogenesis of culture, as shown by the enormous variety of different cultures. Independent of whether in itself it was adaptive or not, every cultural trait could survive that had not damaged the replication of the given culture. A great number of cultural traits having neutral or small harmful effects may have been fixed. Competitive selection was mainly negative because it selected against traits that weakened the ability to compete with different cultures.

The emergence of allopoietic cultural machines and further phases of identical replication

The development of group societies and their identical replication have led to a stable equilibrium lasting for several hundred thousand years. These most developed compartments of cultural autogenesis could have provided building blocks for an even higher organization. Stable group societies were capable not only of replication in time but of replication in space, and as a consequence separate groups speaking the same language and having the same customs appeared. Group societies could either cooperate or be hostile. It can be assumed that at the beginning of linguistic isolation conditions were more favorable for hostility because the groups were small and scarcely capable of vigorous growth. Later in their evolution, when they had already dominated their environment, they often divided, and common language and customs became *bonds* between "daughter groups." Some kinds of cooperation might follow; the exchange of artifacts, ideas, and women became the forms of their interaction. Such a period was revealed by archaeological finds that uncovered traces of mass huntings with highly developed hunting techniques. (See, for example, the analysis of a Central American find by MacNeish 1972.) In locations suitable for beating game, traces of large butcheries (bones of thousands of killed animals) and traps built by many people were found. Such a hunting technique requires the close cooperation of several large groups, elaborate action plans, rigorous discipline, and a hierarchical order similar to that of modern armies. Since the taking of such an enormous mass of prey far exceeds the amount participating groups need for their replication in time, it could have been done for sport, to serve cultic aims, for trade, or for mere enjoyment, although to obtain direct evidence is almost impossible. It is very unlikely that small groups were able to perform such feats. The most plausible explanation is the emergence of a new cultural compartment. Man began to learn the organization of groups permanently cooperating for a given aim and accepting discipline; i.e., *allopoietic machines* emerged. This organization did not simply mean that more people were available for a certain activity, but that participants were willing to submit to the necessary training and discipline and to accept leaders who were able to organize the entire action. Development of allopoietic cultural machines made the integration of strangers for a given function possible, and as a consequence the particular cultures could spread beyond an originating group—for example, as in the culture of taking and exploiting slaves.

The new social organization—groups with allopoietic machines—could cooperate and produce surplus to be used for purposes beyond mere existence. While in group society we find many organizational elements that also are found among animals, organizations similar to allopoietic cultural machines are not known in the animal world.

At the end of the last glacial period a large part of the Pleistocene megafauna became extinct, and much evidence exists that man had a major role in their extinction (Harris 1977), most probably by overhunting with the help of "hunting cultural machines." As a result of an advanced hunting technology, this first ecological catastrophe led to a diminished number of large animals, and it also played a role in the development of agriculture and animal husbandry. These are technological processes that require the contribution of disciplined groups, and they provide surplus beyond the needs of producers. Agricultural production is the consequence of special allopoietic cultural machines.

The "neolithic revolution" can be characterized by the rapid emergence of new technologies—for example, the domestication of animals in a historically short period of a few thousand years, in most cases by capture of whole herds of wild animals (Bökönyi 1976). Therefore, domestication is more an invention than a gradual development. Invention in this case means the organization of a technological process, that of a cultural machine carrying out animal breeding.

The formation of new cultural machines was accompanied by the appearance of a new social organization, which may be called "village society." The formation of a village society requires permanent dwellings, the division of work, and the cooperation of large communities. It seems not very farfetched to compare the differences between group society and village society to the difference between prokaryote and eukaryote organization. In both cases an early, primitive organization changes into a more complex entity, resulting in a new building block of a higher organizational level. As we can recognize modified prokaryotes as cell organelles serving special functions subordinated to the metabolism of their eukaryote hosts, thus we can recognize the ancient group societies as the "special organs" of the village society, such as groups of merchants, craftsmen, tillers, and herdsmen.

It must be noted that among various cultural machines, "fighting machines" also have appeared. We know that hunter-gatherer group societies fought one another. Conflicts were part of their normal life and were connected with territorial defense. However, these conflicts cannot be regarded as wars. Wars began with the emergence of specially trained, disciplined warrior groups, and with the aim of taking slaves, loot, territory, etc. We already have mentioned Mumford's (1967) analysis that was the first to consider these organized groups of people as "machines." The appearance of fighting cultural machines is the most important feature of the interactions of village societies. Perhaps it is the very root of the emergence of early states in which the village societies found some defense and possibilities for cooperation, so that they could remain relatively intact compartments within them. Inputs and outputs of the cultural machines can be joined into a *superorganization*. Therefore, allopoietic cultural machines represent the system precursors of the states that are the compartments of a next organizational level.

During the development of states the allopoietic cultural machines underwent further perfecting. New technologies and new services were invented; common *mechanical machines* also were invented that were carriers of an idea complex, in that they served particular purposes, could produce surplus, replicate artifacts, etc. Later in the industrial revolution, from common machines and humans cultural machines of a new kind were created, and modern industry and technology emerged (Ellul 1965).

A comprehensive and complete survey of societal evolution based on the autogenetic model requires more effort and space than we have. Only *technical evolution*, a particular consequence of the emergence of industry, will be discussed to illustrate the application of the autogenetic model.

Technical evolution

The zero-system and phases of replication of technical evolution

The early phases of cultural and technical evolution coincide; the zero-system of technical evolution is the product of cultural evolution. In the course of cultural evolution a "creative technical-space" evolves, that is, an artifact-producing mechanism is acting.

Artifacts can be created in different ways; their shape and structure can be modeled after natural objects or can be invented by exploiting the human mind's displacement ability. Usually they are produced by *copying* and partially changing an already existing object. Their production, therefore, can be regarded as a *replicative process* in which the accuracy of replication may vary between the wide limits of 0 and 1. The structure of man-made objects lends itself to mutation and recombination; their production is accompanied by selection resulting from competition for raw materials, energy, and production capacity. Thus the artifacts are structures capable of evolution.

In the early phase of artifact production, only very rough copying takes place, and the objects themselves are less differentiated (Pattee 1973). The production of objects without machines is piece production; the "master" individually forms the material and produces the artifacts. The multiplication of artifacts is slow, and the accuracy of replication is low. In time artifacts gain functions and the ability to influence one another's probability of genesis, whereby their evolution accelerates. The function of artifacts used as tools to produce other artifacts is a qualitative difference between the use of objects by man and animals. Animals never use objects to produce other objects. With the appearance of functions, the process of differentiation begins: individual objects prepared for special purposes are produced. As the objects proliferate, different series of objects coordinated on a functional basis appear (such as the special tools for leather preparation). These series of tools are actually supercycles, that is, component complexes influencing in a regulated manner one another's probability of genesis. The increasing replicative accuracy of artifacts is accompanied by a high degree of variability. Obviously, only objects reproducible with great fidelity can differentiate.

Parallel with the appearance of machinery and industrial production, the identical replication of artifacts begins. Quasi-identical objects are prepared by the millions. Nevertheless, random changes keep playing a significant role in the evolution of artifacts. Sahal (1976) has pointed out that innovation in airplane construction happens completely at random, like mutations, and not in the direction generally expected. In particular, parts are developed that are not necessarily those most important for the entire construction. The number of different artifacts suddenly grows, differentiation connected with identical replication increases, and large supercycles (cultural machines) appear. *Compartmentalization* begins; the computer and its peripheries, for example, can be considered a separate compartment.

The separation of technical and cultural evolution

The autonomy of technical evolution might be challenged on the basis that the objects represent only the cultural ideas acting in their production, and thus under all circumstances they are subordinate to the process of cultural evolution. It can be shown, however, that technical evolution is presently in the process of separating from cultural evolution.

This separation is accomplished in three phases. The *first phase* is the creation of the artifact itself. The ideas of cultural evolution, embodying the concept components existing in the human brain, are themselves physical realities. After being made, the artifact is separated from the mental template and may outlive it by even a hundred thousand years, expressing its function and interrelations independently of the mental template. In this first phase, however, the mental template at least plays a role in the object's birth. Thus the separation is only relative.

The *second phase* of separation starts with the appearance of machines, especially automatically controlled machines. After one pattern is established, millions of copies can be made; the ratio of the number of copies made to the number of mental templates involved in their production increases enormously. The mental template of the silicon chip of a pocket calculator used in millions of copies was available perhaps in one or two copies before production. The individual using a complex object is not at all aware of the fine structural details of the artifact in hand.

In the *third phase* of separation the mental template may be completely eliminated from the process of artifact production. Obviously, we are now on the verge of this process. Computerized production planning and the use of robots in production will soon reach the level where the replication of artifacts becomes a self-maintaining process that will not require human cooperation.

This process will be accelerated by the creation of artificial intelligence. Piaget's scheme for the development of human intelligence has to be supplemented at this point. To the observational categories of self-self, self-object, self-(object-object), a new one is added in which the relationship between objects is observed or controlled by an "artificial self." In the course of this process of separation, complex industrial processes, acting without human interference, will certainly develop with a mode of organization that will approach self-replication. Thus there is the possibility of *convergence* in future technical evolution, but we still are quite far from this today since the closed cycles of materials used in technology, being prerequisites of convergence, occur only sporadically and in an imperfect state. In the final phase, however, a global, self-sustaining, continually replicating technological system may develop through convergence.

The parametric information content of the structure of artifacts consists of those parameters of the materials, method of production, and shape of the object that have not been brought about under the influence of other objects. A stone or wooden tool might have a high parametric information content. A rapidly corroding (thus under natural conditions nonexistent) metal object has a high functional information content. That part of the functional information content involved in the replication of the artifact, i.e., the replicative information, also can be recognized. It is easy to see that this part continually increases in the course of technical evolution until the whole system becomes one replicative unit.

Autogenesis of the global system

The highest organizational level of the evolutionary process on Earth is the *global level*, which, as a final level, has some special characteristics. First of all, it consists of only a single entity, a single compartment—that is, the *global biocultural system*. The whole system is a component system. Its components are the various human societies—group and village societies and states, including components of the lower biosphere levels. Components of the global system were continuously assembled and disassembled during human history under the influence of the Earth's energy flows. The present state of cultural evolution seems transient. The highest organizational units are states that ensure the functional coordination and replication of various cultural machines. The dimensions of the controlling forces, or constraints, of the states have reached and even exceeded the size of Earth. Mass media such as radio, television, and communication satellites make communication among people direct, quick, and cheap, but at the same time they promote the manipulation of people by special groups; they also make their own manipulation by other states possible. Enormous amounts of material and energy flow among the states, and there also is a large migration of people. Devices of modern warfare can be directed from any point of Earth to any other point, by which it is possible to destroy not only states but whole continents—or even the whole biosphere.

Open societies operating with allopoietic machines are interconnected, and their interactions may lead to the formation of a higher organization (figure 23). As we have seen, stable component systems are characterized by continuous replication of their components as well as by close coordination among their replicative processes. States on Earth do not yet meet these requirements. Instability is shown by exponential population growth and by uncontrolled production of very effective ideas and artifacts, i.e., uncontrolled production of their main components. Instability also is evident in the fast destruction of the biosphere and in deep social and political crises.

In the former applications of the autogenetic model it was found that if on a given level of organization replicative compartments influencing the probability of one another's genesis develop, then formation of a new organizational level starts and at the same time the

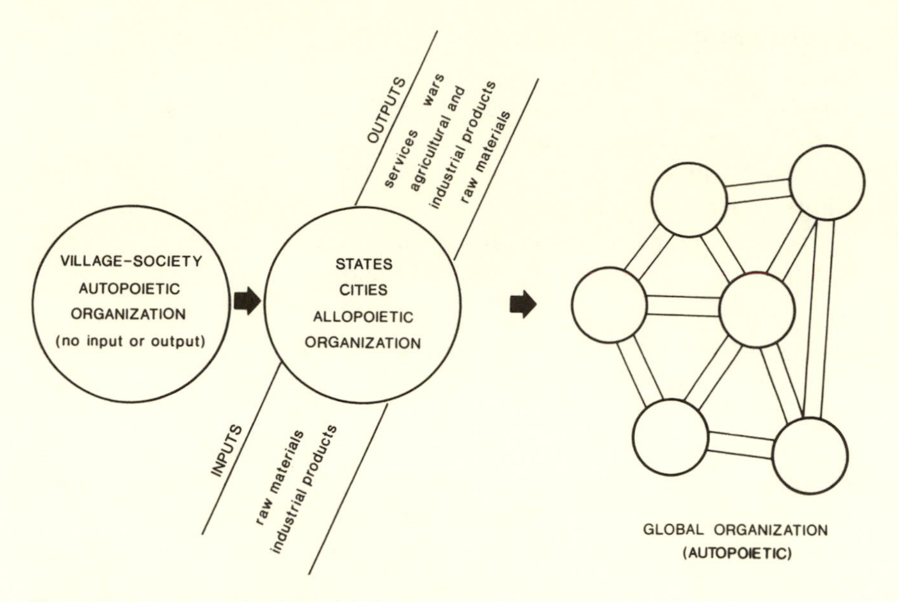

Figure 23 Autogenesis of the global system.

former stability ceases. Conditions for a more or less stable replication of local biosocial systems were provided by the emergence of states during historical times. The early Egyptian, Indian, and Chinese states have been stable for thousands of years, although they have not lived in peace. As we know, replicative systems can replicate in time and space. Besides replicating in time, states can replicate in space while there is free space available. They can conquer territories and colonize them by sending people and ideas, often using force.

Presently the possibilities for replication in space are practically nonexistent. Cultural constraints have emerged that do not allow for the complete destruction of conquered territory and the elimination of its inhabitants. At the same time the states have the power to effectively influence one another's replicative processes. Further evolution of the biosocial system may continue only if new regulatory mechanisms are created that help to form the coordinated *temporal replication of the whole global system*. As autogenesis continues, fidelity of replication as well as stability of the system's components increase. But with the formation of the higher global organization, significant parts of their autonomy will be lost. In a final steady state the global system and its components will be replicated with high fidelity, and its continued existence will depend entirely on constant outside (cosmic) conditions.

The formation of the global system already has been recognized, and several global models have been worked out (Meadows et al. 1982, Richardson 1984). Recognition of global problems will certainly spread, and greater efforts to solve them will be made. These new problems cannot be understood or solved in the framework of former social paradigms. Applying a new system theory is necessary, and this theory basically has already been worked out (Laszlo, 1972, 1973, 1978, 1986). The development of new learning mechanisms is inevitable if we are to transform the idea systems of present societies (Bánáthy 1973).

8 Fundamental Laws of the Evolutionary Process

In earlier chapters we have scrutinized those real systems that exhibit evolutionary processes. Now we will attempt to summarize the general laws, admitting that the proofs of the various theses are often insufficient or are occasionally replaced by mere assumptions. The concept of evolution is applied generally to the dynamics of the whole system, that of autogenesis to the particular cases.

The theorem of matter and force

During evolution that takes place in a physical system, components assemble and disassemble through time. As time progresses, the proportion of matter actively participating in the process increases, and the magnitude of forces influencing the interactions of the components is enhanced until its growth reaches the system's physical limits.

The theorem of energy

In the course of evolution the system moves away from thermodynamic equilibrium and its entropy decreases, while the amount of energy stored, the time spent by that energy in the system, and the organized complexity of the system increase.

The theorem of information

In the course of evolution simple organized system precursors spontaneously appear, and by an autogenetic process they begin to replicate in time. Thereby the replicative information content increases in the system. This replicative information undergoes compartmentalization and converges until the whole system constitutes one replicative unit. At this point the system is in a state of evolutionary equilibrium; its internal control and stability are maximal, whereas its stability toward changes in external factors is minimal.

The levels of evolution

At the beginning of evolution the size of the compartments organizing the coordinated replication of components is determined by the building blocks of these components. As the first compartmentalization is attained in an evolutionary system, the compartment as a new unit itself becomes a building block. If other conditions of an autogenetic zero-system also are present, a *new level* of evolution begins, implying all the phases and phenomena described by the laws of evolution. The serial compartmentalization and development of new evolutionary levels continue until a unified replicative system develops in the whole physical space of the system, and each structural unit of this system replicates in full harmony and with equal probability with all others. This is the state of *evolutionary equilibrium*. On the basis of real systems, the levels of evolution on Earth are as follows:

1. Molecular		
2. Cellular	2a. Neural	
3. Organismic	3a. Cultural	3b. Technical
4. Ecological		
5. Global		

The direction of evolution

The direction of evolution originating at the zero-system points toward the maximization of replicative information. This follows through the convergence of compartments according to the hierarchy of evolutionary levels. At the successive evolutionary levels the physical space of individual compartments increases in size. On this basis, the development of evolution on Earth moves toward a uniformly replicating global system coordinated at all evolutionary levels.

Note

Theories of evolution deal only with system parameters and can predict only the system's general features; they cannot be regarded as deterministic theories. Evolution on Earth—and within this evolution the history of human societies—is a *unique story*, singular and irreproducible. There is no contradiction between the predictability of changes in the general system parameters and the unpredictable nature of the particular forms manifested by the system.

Humans, as replicating beings, are a product of evolution and are subject to its general laws, but humans can choose their own path of history, and this comprises their *freedom*.

9 The Problems of a Finite Earth

The biosphere

Relationships between the biosphere and group or village societies—even early states—were very simple. The biosphere was the *environment*, an unlimited source of food and raw materials, to the societies. Effects of man on the biosphere were negligible; his actions did not disturb the enormous energy and material fluxes of the living world. This situation has gradually changed, although not so suddenly as we tend to suppose. The extinction of several species during the Pleistocene as a result of overhunting has been mentioned. Man increasingly influenced the biosphere by inventing cultural machines, and at present he is the most active agent in the biosphere. It is very unfortunate that recognition of the active, dynamic nature of the biosphere came so late. Until recent years it was seen as a rather passive, static, resourceful environment, subject to all possible exploitations. Recent data have shown that it is the biosphere's natural regulatory processes that provide and maintain the conditions for life, e.g., free oxygen, reasonable temperature and humidity, and many other important parameters.

Our most urgent problem is that we do not know enough about the biosphere's regulatory processes. Presently we use infinitely simple physical models that have a very limited predictive power. We do not have the necessary theoretical basis to deal with systems of this complexity (Botkin 1982).

The most apparent effect of man on the biosphere is the extermination of species and communities, a radical reduction of the biosphere's complexity. Data gathered in the 1970s showed that about 25,000 plant species (IUCN 1978) and more than a *thousand* vertebrate animals (IUCN 1975) are on the verge of extinction. According to conservative estimates, by the turn of the century another 0.5 to 1 million species will become extinct (Myers 1979). Interpreting the dangers of this decreasing complexity to the public often takes the form of pointing out that many effective drugs or new kinds of food originate from plant and animal species that have been considered useless. This is a very simple argument; heretofore, it is as if the biosphere were nothing more than a resource for man, and the only reasons for conserving it were for man's future use. The fact that the biosphere is a *system* of which we are components, and the fact that we are ignorant of its complexity and do not know the extent to which we may modify it without destroying it, has been realized only by a minority of human societies. It also is quite possible that the complexity of the biosphere has been evolved not by mere chance, and that its permanent stability depends on this very complexity (Allen 1980). Domesticated species propagated in enormous numbers may temporarily be extremely useful for man, but it may happen that the species that were pushed out by the monocultures have played important roles in supporting our basic life conditions. For example, we know that in *desertification* human influences have been and still are the most important factor (Spooner and Mann 1982).

Besides the decreasing complexity of the biosphere, pollution is the most important human influence, and it is not at all independent from the former. By unlimited burning of fossil fuels, the carbon dioxide, nitrogen oxides, and sulfur content of the global atmosphere is continuously increased. Some of the harmful effects are local, but often they spread over continents as in the case of acid rain. Through these influences, man has become a main formative force of the biogeochemical cycles of Earth, which control the fluxes of the most important biological elements—carbon, nitrogen, phosphorus, and sulfur (Bolin 1981). The flow of these elements is the most fundamental process in the biosphere. Even small changes may cause unforeseen consequences for developed societies, which certainly could not cope with the sudden decrease of agricultural production resulting from a mild change of climate, although the danger of such a change is very real (Ward and Dubos 1972).

So far we have mentioned interferences that are the consequences of normally functioning human societies, but we may add to them the threat of nuclear war. According to the latest studies, a full-scale nuclear war, besides eradicating human societies, might cause the sudden change of climate by cooling the atmosphere for six months to a year. The "nuclear winter" would probably change the biosphere irreversibly and to an extreme extent (Turco et al. 1983, Covey et al. 1984).

In the present situation it seems unrealistic to expect conservation of the still existing natural ecosystems or a radical decrease in the harmful effects of human societies. It is quite unlikely that states would accept rules to halt further damage (e.g., stop eradicating woodlands, control burning fossil fuels by an internationally accepted quota system, stop air and water pollution, etc.). The cause of this pessimism is that acceptance of such international control would cost an enormous amount of money and is against the local and immediate interests of states, while its favorable effects would appear only globally, slowly, and in the future. At present there is no effective and generally recognized organization that could represent or even articulate global interests.

Therefore, it is very important to show how the relations of biosphere and societies have changed radically during the last period of cultural evolution. The biosphere is not the environment of society but is an *integral part of society*. Society's increased influence has been so sudden and great that ecological conditions necessary for normal operation of societies do not seem to be provided spontaneously by the biosphere. Very soon, probably around the turn of this century, we have to control and provide the very conditions of our own ecological existence. Whether we want it or not, we have to take responsibility for certain aspects of the biosphere's control processes. As a consequence, society and the biosphere will merge into a *unified global component system*, consisting of humans, ideas, artifacts, and all living organisms as components. Instead of spontaneous control processes, the arising system will rely on artificial, technical control. As artificially controlled agricultural ecosystems replaced natural ecosystems in Europe, the biosphere will be replaced by a socially controlled bioculture. This will happen not because anybody wishes to take over control, but because frequent local catastrophes will compel society to take control, whatever it costs.

Population growth

Human population is growing exponentially, and at the year 2000 it certainly will exceed 6 billion, loading the carrying capacity of the planet beyond imagination. In less-developed societies some form of birth control always existed, in many instances without an awareness of the people who practiced it. Stable replication in time requires a constant population to maintain equilibrium with its environment. At present the rate of population growth is inversely related to society's development. Whether the population of a given society is optimal or not is a difficult question, considering the quality of life, but we cannot go into details of this problem here. The population problems of the poor countries have mostly resulted from the rapid decrease in mortality rates. There were improvements in health conditions and medical care, and some powerful new drugs were introduced, etc. Unfortunately, these changes were not inherent consequences of organic development of the given societies but were almost always imported from more developed countries in the form of ideas, technologies, or active materials. An unforeseen and unwanted consequence of these processes is that the spread of the new ideas, technology, etc., has not been accompanied by the appropriate, supporting cultural changes, which, for example, would control birthrates on a level corresponding to new societal conditions. Here is a clear example of an *external influence* on the replicative process of the most important component of the human society—man himself.

In a stable society replication of human beings does not simply mean the birth and survival of the individual but also the replication of those societal conditions necessary for a decent life in a given society. If replication of a component in the component network of the society speeds up or slows down, the stability of the system decreases. The situation becomes worse when the natural population control mechanisms are suppressed by an otherwise very respectable humanitarian aid of food, drugs, and medical assistance that helps to avoid the immediate catastrophes but prevents the operation of the cruel correcting mechanisms. As a consequence, the situation becomes more catastrophic in the long run. Of course, it is inconceivable that the developed nations should let the overpopulated poor ones tackle their own problems and allow the eradication of people by famine or plague (the control processes of nature). It is impossible because very effective idea complexes were formed in the developed nations that extend the concept of human community to the populace of the whole planet. This notion seems to be trivial nowadays, but in primitive societies very frequently the concept of "man" has meant a member of the tribe, and humans belonging to other tribes were not included at all. There were times when ethical laws were not extended to strangers; they could be killed, tortured, or enslaved. Unfortunately, the new extended humanitarian ideas are not yet effective enough to create a global organization that could solve the population problem.

To solve such problems, recognition of the necessity to subordinate particular interests—whether they be individual or state—to global stability is essential. The exponential growth of population is threatening the depletion of Earth's carrying capacity. Although crises are local in the short run, they soon will spread to the entire planet. Developed nations must recognize that population problems have been caused by their intervention in the replication processes of other societies, so now they must help them solve these problems, which certainly involves much more than respectable humanitarian aid that gives only temporary relief. Perhaps it also is an important argument that finding a solution also is in the utmost interest of the developed nations.

Economic problems

The past few decades have been characterized by a deepening economic crisis. Recessions, the oil crisis, inflation, and the enormous debts of various states are carrying the world to a global crisis and a possible collapse of the world economy (Brandt Commission 1983).

According to the replicative model, these problems are connected with the emergence of the global organizational level. As has been discussed, the societal entities of early cultural evolution, the group and village societies but also the early states, were almost closed economies. They replicated almost perfectly for centuries, and their components of humans, artifacts, and ideas composed a closed creative space. In the next period some societies developed allopoietic cultural machines and have never attained equilibrium because the outputs of their cultural machines were more than necessary for their needs, and they were able to influence the processes of other societies. Surpluses that were created by the cultural machines made trade, exchange, and mutual economic interactions possible. Aside from war, economic interactions are those forces by which states are able to influence the probable genesis of each other; that is, those forces will lead to integration and the creation of a higher organizational level.

The artifacts produced by cultural machines have to be functionally distributed, and material and energy inputs for operating them should be provided. But these are the problems of the more-developed industrial systems. When these new ideas and new artifacts appear in the creative space of the underdeveloped states, they cause disturbances in the stability of their replication, which is the central problem of the underdeveloped nations. If such new artifacts are able to join with the existing replicative network of functionally connected components, i.e., if they have functions in the framework of the colonized system, then they begin to influence the probability of genesis of other components. Some will increase, while others will decrease or completely perish. This process may disturb the colonized system because it is not always capable of *replicating components coming from outside*. The higher the difference between the levels of development of the two interacting systems, the greater is the possibility of this replicative failure. As a consequence, the given artifacts do not become components of the system but remain *outside conditions* for them. If the system adapts to these outer conditions, that is, the necessary changes occur in its component network, then it becomes dependent on the conditions represented by the artifacts. If for any reason the continuous supply of the given artifacts is interrupted, serious disturbances are immediately created in the system. For example, let us suppose that state B provides a large number of tractors to state C, which has not previously used such machines in its agriculture. The presence of the tractors obviously will change the agricultural technology, but at the same time it will make the poor country dependent on the continuous supply of these machines. Tractors must be produced by state C, which is not a simple task because to produce tractors many other artifacts are necessary, most of which must also be imported from outside. And thus the problems multiply. Or state C needs continuous import of the machines themselves, which will cause it to lose its economic independence.

It seems logical that these imports can be financed by increased agricultural production, but this is not so because usually not just a single artifact but whole technologies, whole cultural machines, are involved, the effect of which is very wide and penetrating. Fields plowed by tractors will be more productive only if fertilizers, insecticides, and harvesting machines also are introduced. These new artifacts, in turn, create ever newer problems since they all appear as outer conditions. It must be added that not only are artifacts involved but also the idea systems expressed, either in the artifacts themselves or simply accompanying the artifacts to regulate their proper use, operation, repair, etc. Some of them will not be introduced into the new system, which will cause further problems.

Of course, it is possible that an effective artifact can be introduced slowly enough to allow the accumulation of the replicative information of the artifacts in the underdeveloped system and thus avoid harmful effects. If the introduction is too rapid, then the less-developed economy becomes simply integrated into the more-developed one's replicative network by consuming some of its products. By this integration a series of transformations begin that will change the whole network of the underdeveloped state, for usually the ruling group is not aware or does not care and delays the necessary social and political adjustments.

At present avoiding the influence of developed states is possible only by complete isolation, which is possible only for a few countries (e.g., Albania). The process of integration, therefore, is going on with all its bad and good effects. Its harmful effects can be mitigated if the economy's replicative nature and the real nature of the interactions among the states are recognized. Gradual development of underdeveloped states respecting their own replicative processes and providing not primarily technology but *replicative information* (know-how) would allow these states to develop their own balanced and proportional replicative economies.

Mutual interaction may also cause problems in the more-developed states. The replicative network of the more-developed state also is adapting to the situation in which the output of its own cultural machines is not fed back to its own replicative network. Therefore, when the situation is changing, e.g., the demand for products diminishes, then overproduction causes problems in the developed economy. The states usually help their economies by shifting the harmful effects of overproduction toward other states, thereby ensuring the smooth operation of their cultural machines.

The solution to these very difficult problems will be possible when the integrative process of the global system is recognized, accompanied by acceptance of the concept that not only the production and distribution of artifacts require control, but several political and social problems must be solved in order to make the formation of a global society possible. Instead of forecasting various catastrophe scenarios, it would be wiser to search for positive solutions.

Appendix 1

The evolutionary theory that has been presented here is a generalization of Darwin's theory (Darwin 1859). It has been formulated at an intuitive level, although it seems to lend itself to a more axiomatic formulation similar to the formulation of the Darwinian basic theory by Williams (1970) (see an attempt by Kampis in appendix 3). Nevertheless, despite the limitations imposed by its character, this theory claims to support the notion that evolutionary capability is an *inherent property of matter* that can be traced back to simple physical laws. As a sufficient number of simple molecules at a properly low temperature compose an open system and are exposed to an energy flow capable of exciting the molecules, the system immediately moves in the direction of higher organization. Organized, simple system precursors spontaneously appear, and the development and concentration of replicative information begins. Obviously, the dimensions and physical properties of the system influence the constituent processes—compartmentalization, the number of individual evolutionary levels, etc.—but these are matters of detail. Given the necessary time, the system reaches a terminal state in which replicative information content is maximal, processes are coordinated, and further changes cannot be initiated *from within*. As long as the parameters of external energy flow are stable, the system as a unit exists in the steady state of evolutionary equilibrium. The final state is contradictory, for the internal stability of the system is maximal, and at the same time it entirely depends on external conditions, as it has completely adapted itself to external circumstances during the long process of convergence. If the system reaching evolutionary equilibrium is part of a greater system (compartment), and its physical boundaries are narrower than those of the whole system, the next level of evolution appears through subsequent replications of the recently assembled replicative unit, and the accumulation of replicative information continues. If the compartmentalization and convergence of replicative information have reached the physical limits of the total system, a final state of equilibrium is reached. Now the total system (itself a replicative unit) is ready for further replication, but the necessary matter and physical space are not available. Its further existence, therefore, is an eternal process of self-renewal, a replication within itself.

Presently we are in the phase of evolution on Earth when the signs of a final global convergence are recognizable. The question arises whether evolution is a phenomenon limited to Earth, or possibly to the planets, or whether it has a role in shaping the fate of the entire Universe.

My own, definitely wild speculations can be expressed in a generally unjustifiable, logical explanation as follows: If the global system reaches the steady state of evolutionary equilibrium, it becomes a replicative unit. If, as a consequence, by colonization from Earth, or independently, further global systems evolve on other planets (see Kupier and Morris 1977), the next question is whether these other systems will be able to develop *functions*. If the answer is no, evolution is only an interesting but particular phenomenon of the physics of lower temperatures restricted to a minute part of the Universe. If the answer is yes, a new evolutionary level appears, and further evolution will influence the parameters of the galactic system. Although it appears that available data are still insufficient to calculate the terminal state of the Universe (Barrow and Tipler 1978), on the basis of this general theory it may be taken for granted that the metagalactic growth of replicative information necessarily leads to the system's increasing physical compactness. In the final state of the total amount of matter of the Universe carries replicative information, it is accommodated in the smallest possible space, its entropy content is zero, its internal stability is maximal, its external stability is zero, and its existence is timeless self-renewal. In other words, it may explode at any moment in a next Big Bang.

The general theory of evolution predicts an oscillating Universe with the reassuring conclusion that the evolutionary *history* of each expansion and contraction is different from the others.

Appendix 2 Considerations for the Scientific Formulation of Epistemology

Epistemology traditionally belongs to the realm of philosophy, but in the course of rapid development and increasing integration of the natural sciences several questions directly connected with scientific practice have arisen. For want of a better alternative, these must be dealt with from within the natural sciences themselves. The most pressing of these problems appear to be in the fields of genetics (Campbell 1960), psychology (Piaget 1967), and ethology (Lorenz 1973).

Rather than discussing individual questions here, an outline of an epistemology is given with the intention of presenting several considerations from the natural sciences that are relevant to this problem.

1. Definitions of cognition and knowledge

First, the meaning of a few frequently used terms should be made clear. We are going to deal here with *systems* that can be distinguished from their environment by their physical parameters. Physical *components* exist within these systems, and they can be unambiguously described by the arrangement of their atoms, which form the structure of the components. The interactions that may occur between such components resulting in structural changes will be referred to as *functions*. As a trivial example, the cell may be considered a system containing various molecular components. It can be described by these components and their interactions, and in turn these interactions can be characterized as special functions, e.g., enzyme functions. An enzyme function always manifests itself in the altered structure of other molecules. When the discussion of the "system" is taken to a higher level—to the level of organism, for example—the cell itself might then be considered a component.

Cognition (which we shall label A) is a function, an interrelationship, that within the given system increases the probability of survival of the component to which it belongs. That is, it increases the chances of survival of the participant actively involved in cognition.

Knowledge (labeled B) is that part of a component that has evolved in the course of cognition. It is a special structural arrangement, the base on which depend the functional connections among the various parts, components, and parameters of the system to which it belongs and the increased probability of survival of the subject (component) involved in cognitive acts.

It follows directly from this definition that a *cognitive process* is an event in the course of which knowledge is acquired. The first two definitions (A and B) almost will certainly evoke controversy; for this reason it must be clearly stated that the present definition of cognition must be considered within its present area of validity. Despite these limitations, the above definitions can offer several advantages, namely, they help us arrive at general principles by virtue of which the cognitive processes of humans (including their evolution) as well as lower "biological" cognitive processes, or even the possible cognitive processes of technical constructions (by no means a negligible question nowadays), can be uniformly interpreted.

2. Levels of cognition

Although the above definitions refer to cognitive processes of individual components and subsystems, they also can be interpreted on a systemic level with minimal supplementation. Individual cognitive processes take place within a given system, and the acquisition of knowledge alters the structure of the cognitive components. As a consequence, however, the system itself changes because it acquires new (transformed) components in the cognitive process. If such a change increases the given system's probability of survival and reproduction, then a cognitive process also may be regarded as having taken place at the systemic level, and the resulting "knowledge"—in the form of altered components—as defined can be attributed to the total system. This also follows from the definitions of system and component, since any given system can be thought of as a component at a higher level.

3. The mode of cognition and the content of knowledge

In these definitions only the most essential features of cognition and knowledge have been established. However, further questions concerning the content, meaning, and mode of acquisition of knowledge must be addressed. From observing concrete processes it appears that the structure corresponding to the knowledge of a given component is always a construction reflecting in a simplified form some feature or relation of the environment—that is, the subject of the component's cognition. Thus knowledge is always a model (Mesarovic 1964). Consequently, the mode of cognition consists of the *construction of a model* from concrete, available structural elements with the ability to reflect, more or less accurately, one or more aspects of the environment. Such a definition also can be readily justified intuitively, as knowledge indicates a certain kind of orientation and foresight. This feature manifests itself exactly in the capacity to build a functioning model and to use the data provided by the model for prediction.

4. Theory of multiple reflection

The model theory of knowledge is widely accepted and is the basis of Marx's reflection theory. However, it can be extended by one essential, logical step. If a component is capable of cognition, and in the course of acquiring knowledge it reflects its environment in a model, this process certainly will affect the environment that contains the cognitive component. This is obvious, since according to our definition of cognition the probability of the cognitive component's survival increases as a consequence of reflection, with consequent changes in the higher systemic level as well. A component's environment always can be viewed as a higher systemic level. Alterations in this higher system level that result from the cognitive process create a new environment for the cognizant component, which in turn produces more knowledge. (But this already contains the previous step, so to speak.) Thus in the long run knowledge-acquiring cycles are of an iterative or feedback nature. A *multiple reflection* of the environment takes place in the cognizant component. This can be formulated in the following thesis: The acquisition of knowledge (labeled C) alters the higher systemic level (the environment) in which a cognizant component acts, and this changed system is reflected repeatedly in the cognizant subject. Knowledge acquired by the subject during this process is cumulative.

This phenomenon plays a fundamental role in the history and fate of systems and therefore will be discussed later.

5. Cognition in real systems: the cognitive processes of genetic structures

Before elaborating on some features of the cognitive processes that follow from the previous discussion, it is worth considering some fundamental real systems. The basis for the evolution of living organisms is the gene store of the various species. Genes are definite molecular structures, and the functional relationships of these components to one another and to other components within their environment are the subject of molecular biology and genetics. Another and possibly less obvious feature of the genetic "system" is that, being a cognitive system, it is a model of its environment.

On the surface of Earth the distribution of light is constantly and regularly changing. This has undoubtedly been a basic characteristic of the biosphere for billions of years. Accordingly, the metabolism of most species shows evidence of an inherited internal rhythm encoded in the genome. Different species may be active in periods of darkness or light, but all have the inherited "knowledge" of this environment's periodicity. The genetic "regulation" of this internal rhythm is relatively well known, and certain species also appear to possess a similar knowledge about lunar movements and the changing seasons. These all share the characteristic that, although the actual external factors may change the internal rhythm and may to some extent even override it, the foundations of these rhythms are genetically coded and inherited according to rigid rules.

Let us take another example. In recent years the selection of routes by migrating birds has been thoroughly studied. It has been shown that storks inherit the knowledge of the main parameters of their migratory routes. European storks use one of two paths: birds from the eastern regions fly to Africa via the Bosporus, while "western" storks travel via Gibraltar. If young eastern storks without personal experience of their traditional migratory route are caught and released west of their breeding ground, they start to migrate to the east, that is, in the direction appropriate to the area in which they were born. Western storks behave similarly; when moved east they start to fly west in accord with their place of birth. This suggests that the storks' knowledge of migration routes is rigid. It is not affected by environmental factors but is fixed genetically. Thus storks may be considered to possess a "map" of the migratory route in the genome. However, this does not seem to be true for all species of migratory birds. In the case of the greylag goose such a map cannot be found in the genome but develops instead in the central nervous system through learning (Eibl-Eibesfeldt 1970).

Not only cyclic alterations of parameters, or the mapping of migratory routes, but often very complex "skills" can be found encoded in genetic material. There is a tiny moth that is a parasite of the yucca plant. The female lays her eggs in the flower of the yucca, and the larvae feed on the developing yucca fruit. The larvae can survive only if the yucca flower develops into fruit, that is, if it is fertilized. This vital act is carried out by the female moth herself. She collects pollen from the stamen of the flower, rolls it into a small ball, and places it on the stigma. This process of fertilization is a complex "technology," carried out by the moth merely on the basis of its inherited knowledge, and is encoded in the genome (Morgan 1896).

The mechanism underlying the acquisition of knowledge by genetic components is known. Random alterations of structures, e.g., mutations, produce variations in the components. As a result of environmental influences, the components that become established are those that can best be incorporated into the already existing system's functional network. Such genetic cognition is a relatively slow process, for the acquisition of essential new knowledge sometimes requires millions of generations. It is characteristic of the genetic accumulation of knowledge that in the course of its development the proportion of knowledge about the cognitive component itself continues growing. In the course of evolution the genetic systems of populations of living organisms develop an ever-increasing knowledge concerning other functionally related populations. If the gene store of the whole biosphere is considered as one system, and this system's cognitive processes are studied, it is possible to demonstrate a tendency to move toward a steady state, as is the case for all systems composed of physical structures. At the same time, and as a consequence of the cognitive processes occurring in the system, the equilibrium of the system is in disintegration. The very knowledge that corresponds to a newly emerged steady state disturbs this state. The acquisition of knowledge thereby becomes the driving force for continuous change.

6. Cognition in real systems: neural cognitive structures

In the first long period of evolution, only the genetic level of cognition had come into being. It was only with the development of animals' nervous systems that a fundamentally new mode of cognition appeared.

Each species reacts to stimuli received from the environment in a characteristic way, with the most primitive form of response being a direct reaction (Lorenz 1973). Many reactions of unicellular organisms belong to this category. Organism and environment affect each other simultaneously in this case. With the development of multicellular organisms, a new type of action, *sensitization*, appears. In sensitization an excitation received and transmitted by the receptor cells produces in the nervous system not only momentary, transient reactions but lasting states of excitation. Should the stimulus reappear, the organism's reaction becomes faster, more direct, and more effective.

The mechanism of *habituation* is similar. Experiments with marine mollusks show that a habituation to touch results from the exhaustion of post-receptor interneurons. The possibility of habituation makes the "disconnection" of repetitive stimuli (that is, reserve and inactivity) possible. These complementary processes provide the basis for higher neural functioning. A triad consisting of *receptor cells*, *interneurons* capable of sensitization and habituation, and *motor neurons* actually make up a primitive environmental model.

The motor structures do not operate as a direct response to stimuli. Rather, the information content of the stimuli first enters the internal model consisting of interneurons, and the resulting activity of the model regulates the motor structures. As long as the interneuron network is relatively simple, this modeling of the external world remains extremely primitive. The model's structure consists of only a few points of excitation. In higher animals, however, very accurate representations of the external world can be demonstrated. The mental model is not simply a reflection of the environment; it also can reproduce the dynamics and functional relationships of external processes. In addition to representing the external world, it contains possible behavioral instructions for the organism (MacKay 1951–52). The animal's activity is not regulated simply by reactions to environmental stimuli, but by the expectations and situational investigations arising from an analysis of the model developed within the nervous system.

This model is the brain's representation of the environment and is used by it as an internal reference for the control of various forms of behavior (Sokolov 1960, Archer 1976, Csányi 1980, 1982, 1985b, 1985c).

According to leading neurobiologists, the modeling activity of the nervous system is the cognitive process itself. Careful analysis of ethological data indicates the existence of cognitive functions in animals as well (Griffin 1976). This view is strongly supported by the fact that the vertebrate brain, especially that of mammals, is very similar in structure to that of man. All the evidence seems to suggest that the mind developed gradually, not in a single leap, and that differences among species are of a quantitative nature. However, concrete physical structures containing neural models are not yet known. They might exist at the neural level, but it is possible that subcellular components play a determining role. Whatever the case, there is no doubt that the nervous system's modeling activity gives rise to structural alterations corresponding to our earlier definitions of cognition, knowledge, and system alteration concerning cognition. The biological benefit of the acquisition of knowledge by animals is the increased probability of the individual's survival. Memory traces play an essential role in the neural model's development. Learning, a process in the course of which memory is established (within genetic limitations), is the mechanism that selects, among model variations formed in the brain by environmental influences, those models that can most likely contribute to the organism's survival. The selected model—that is, *knowledge*—is interrelated functionally with other environmental components, and it is this very functional interdependence of the model that makes the selection of models possible.

The principle of multiple reflection also is valid in cognition at the level of consciousness. Animals having the most advanced nervous systems, apes and man, are capable of building a mental model of *themselves* and thus can be said to possess self-consciousness (Gallup 1970). The structure of the nervous system of every species is genetically determined; the "neural" mode of cognition is a subsystem that has developed above the level of genetic components, and its function is to provide a rapid but transient cognition compared to the genetic system. Mental knowledge possessed by an animal, even that of a self-conscious animal, is generally lost with the individual's death. Consequently, this kind of knowledge as a component does not become permanently built into the system. Its only purpose is to serve the evolution of still better knowledge-acquiring mechanisms, since the reproductive rate will be the greatest for those individuals whose knowledge-acquiring mechanisms under given circumstances are the best. The result of this selective pressure is an increasingly rapid development and consequent appearance of brains that are capable of producing increasingly perfect models of their environment. At the peak of this development is man, with a brain that appears to be in the light of present knowledge not only the most perfect model-building apparatus, but also the first system capable of transcending itself and establishing a completely new level of cognition.

7. Cognition in a real system: cognition above the level of the mind

Numerous essentially biological factors play a role in the development of human society: language, the use of tools, highly developed sociability, and cooperativeness, just to mention the most important. All can be reduced to one essential factor: man is a *structure-building, constructive being*. Language, and the ideas expressed through language, may be regarded as structures, as can such artifacts as tools.

Social institutions based on social relations also may be regarded as structures. Thus, based on his biological characteristics, it seems that man is able to "generate" a new system containing the above-mentioned structures as components, and we call this system society. It is suggested here that these constructions of man are models; that is, they are built up from components containing knowledge. It is further suggested that the new system produced by man is capable of cognition at its own level and fits our definitions of cognition and knowledge. That is to say, man's constructions, whether ideas, artifacts, or institutions, are themselves *cognitive structures* (components) capable of entering into a functional relationship with their environment through the cognitive process and able to build the knowledge thus acquired into their architecture. The knowledge-acquiring process of artifacts is generally slow, the computer being one notable exception. The speed with which it acquires knowledge may eventually reach and even surpass that of the human brain; and it is beyond doubt that computers will be increasingly used in the collection of knowledge independently of human consciousness. Whatever the limitations of present constructions, it is inevitable that those of the future will be capable of gaining knowledge that has not been "fed" into them. At some level of complexity the appearance of self-consciousness is only a question of construction. The largest cognitive structure on Earth at present, both in capacity and operation rate, is human society, composed of biological, mental, material, and social components. The last three have become functionally interactive at extremely high rates, collecting knowledge about the environment and about themselves. This is expressed in the subsequent change of their structure. The accumulation of knowledge has a stabilizing effect on the system's parameters, while the society becomes increasingly self-sufficient.

8. The effect of cognition on higher systems

So far our discussion has centered on the cognitive process and cognitive structures, but now it is necessary to consider the effect of cognition on a higher systemic level in which the cognizant subject acts as a component. These questions recently have received special attention as social cognition causes fundamental changes in man's immediate environment—the biosphere. The direction of change can be deduced from our definitions of cognition, knowledge, and system alteration. As a consequence of cognition, the model of the cognized system or part of the system is incorporated into the cognitive structure. As a consequence of multiple reflection (the definition of system alteration), a model of ever-increasing complexity and accuracy is incorporated into the cognitive structure, with the result that the difference between the real system's functioning and that of its model constantly diminishes, and thus an ever-increasing part of the original system becomes part of the cognitive structure. Since on a certain level of complexity the cognitive structure models itself through multiple reflection, an ever-increasing part of its knowledge refers to itself. Under the combined influences of these two processes, the system and the cognitive structure become better and better coordinated functionally. The cognitive structure of growing complexity plays the dominant role in this coordination.

From these we can postulate: The cognitive structure assimilates the cognized system and becomes one with it (what we label as D).

As for the above-mentioned problem of the biosphere, there is no realistic basis for the notion that it is possible to protect the biosphere from human interference (that is, from assimilation). Society necessarily must assimilate the biosphere by producing new conditions for its functioning. In other words, in the process of acquiring knowledge, the physical mass and complexity of the cognitive structure keep growing to the detriment of the system, until the whole system has become one with the cognitive structure.

9. The limits of cognition

It follows from the above definitions, particularly from (D), that on any given systemic level cognition cannot be *unconstrained*. As a consequence of multiple reflection and assimilation, the system becomes increasingly stable and provides a decreasing amount of information for the cognitive structure—substantially for itself. If the boundaries of the system are fixed and the cognitive structures are unable to cross those and leave the system, and at the same time external factors do not change the parameters of the system by force, then cognition tends asymptotically toward a limit. The system assimilated by the cognitive structure reaches a steady state in which its components lose their capacity to change.

To formulate a thesis (which we call E): In a system that is closed from the point of view of cognition, knowledge accumulates, cognitive processes converge, all components of the system become functionally coordinated to constitute a unified whole, and the system becomes stable and reaches a steady state.

10. Cognition and the universe

Although the various real systems possess numerous special features, the five theses developed above seem to be valid for all of them. Consequently, the phenomenon of cognition is not linked to any special organization or structure. It is suggested here that a concluding thesis (F) can be formulated: Cognition is a fundamental form of the manifestation of matter.

This appears to be valid at least for atomically constituted matter. Cognition is a kind of self-organizing feature of matter that, under proper physical circumstances, appears without further conditions, and through a series of structural changes the properties of the given system converge toward a steady state. The thesis postulates that, as a general principle, the capacity to create structures is "inherent" in matter on any level of development or in any form of its motion.

Whenever a system composed of atoms at a sufficiently low temperature is subject to a flow of energy capable of exciting the atoms, the system immediately moves in the direction of higher cognitive organization. The history of evolution on Earth is a clear example.

11. The place of human cognition in the universe

Man always has striven to find his place in the Universe, often imagining himself to be a unique "creature," a lonely wanderer, and often regarding his mind and talents as exceptional and irreproducible phenomena. Recognition of the rules of cognition supports the notion of man as an inherent part of the Universe, manifesting the general laws of matter in his activity. If man could manage to surpass the earthbound limits of his cognition, the whole Universe would be open to him, with a chance (somewhat greater than zero) that it also could be changed in accordance with his vision.

Presently we are in the phase of evolution on Earth when the signs of assimilation of the entire Global System by man can be recognized. If the Global System reaches a steady state, it has two alternatives. Either its cognitive forces turn on the Global System itself and it becomes a closed system; or it begins to explore its cosmic environment, the Universe, and a new cycle of cognition starts. The New Global System as a cognitive component of the Universe will be an active force and will be capable of assimilating its environment.

Appendix 3 On the More Formal Approach to Autogenesis

György Kampis

Introduction

Since the elaboration of the first version of this theory of evolution (Csányi 1978), and particularly since the formulation of autogenesis (Csányi 1985, Csányi and Kampis 1985), we have been working on a mathematical description and system modeling of evolution (i.e., of autogenesis). Our aim was to eventually develop an abstract, theoretically grounded formal model in which autogenetic processes take place under weak assumptions and which is adequate for the characterization and reproduction of real evolutionary processes.

Certain elements of such a description already have been elaborated, starting with a preliminary description through publications illuminating some problems of the concept of information (Kampis 1986a, 1986b) to studies discussing the concepts of component systems, replication, and evolutionary complexity (Kampis and Csányi 1986a, 1986b, 1986c).

There is no place to go into detail here. The purpose of this summary is only to interpret the concepts of autogenesis more precisely. Thus we shall deal with components, component systems, zero-system, function, replication, information, autogenetic system precursor (AGSP), compartment, etc.

It may be worthwhile to emphasize that this approach confronts the transformations *within* given systems with functions and with replication, and inert information with the information that is correlated with observable actions in the system. We refer to the above-mentioned papers for the background and theoretical importance of these points.

Component systems

Components, built up from simpler building blocks, serve as the basic units of autogenesis. They are constructed according to rules that are uniform but depend on the set of building blocks. The characteristic process of a system of components is that new components are formed from the given ones.

Definition one

Let $E = \{e_i\}$ be the finite, nonempty set of building blocks. The set of (possible) components is the set S, defined recursively by a given $\phi(\cdot,\cdot)$ mapping in the following way: $\phi : S' \times S' \rightarrow S$; $S' = S \cup E$. (Mapping ϕ is called the composition rule.)

The term s_j is the jth type of component (S is obviously countable). If components are formed in the course of the system processes, this always takes place according to the composition rule. Thus it appears implicitly in the transformations, which prescribe only allowed operations on components.

During the functioning of autogenetic component systems building blocks are neither generated nor destroyed. Components are therefore transformed into one another. Now let t denote the (supposedly discrete) time instances, and M_t the state of the system; M_t is the enumeration of components, as a set, and is also the symbol for the set. When saying $s_i \in M_t$ or that M_t is finite, we refer to the proper set. We shall speak of finite systems only; thus there is a finite number of different states: $M^1, M^2, \ldots, M^n$.

Now we can introduce the basic transformations of the system.

Definition two

Let M_t be the state in t. The mapping $f(\cdot,\cdot,\cdot)$ for which (1) $0 \leq f(j,i,M_t) \leq 1$ (2) $P(s_i|_t$ is transformed to $s_j|_{t+1}) = (j,i,M_t)$ for all $s_i \in M_t$, $s_j \in S$ is called the local transformation of the system.

That a component is transformed to another is understood here to mean that the former contains some building block(s) from the latter. Although this does not influence the present inferences, we assume that f is identical for all copies of components of identical types.

The zero-system

The state of the system can be characterized, on the one hand, with the system state taken in the narrow sense and, on the other hand, with its transformations. The set of transformations of components seen at the level of the whole system is denoted by F $(F = \{f\}_S)$; that is, it is the set of $f - s$ which belong to S.

Definition three

The autogenetic zero-system is an ordered triple $O = (S, F, M_1)$, where M_1 is the initial state.

Function

During autogenesis, components appear that change the original f (or F) transformation of the system. This means that while the transformations of the zero-system are independent of its components, those corresponding to functions are component-dependent. It should be noted that there is no possibility for an a priori enumeration of these transformations in some suitable f' since these are realized only with the appearance of the given components and are unknown before that. This is expressed in the definition of functions for individual components, and also for their sets.

Definition four

If $P(s_i|_t$ is transformed to $s_j|_{t+1}) = f_{s_l}(j, i, M_t) \neq f(j, i, M_t)$ for an s_l and some $s_i \in M_t$, $s_j \in S$, then s_l is a functional component in M_t. This is denoted by $s_l \rightarrow s_j$ and $s_l \rightarrow s_i$.

This notation expresses the fact that the function of s_l is related both to s_j and s_i in some way (determined by fs_l). In the sequel we shall suppose that if a component is functional, then it has function for itself as well. This means only that the transformations related to it also depend on it to some extent.

Definition five

The set $F_1 \subseteq M_t$, for which $\forall s_l \in F_1$ is functional in M_t, i.e., $f(j,i,M_t) \neq f_{F_1}(j,i,M_t)$ for some $s_i \in M_t$, $s_j \in S$, is called a set of functional components. The set of s_i-s of this definition is called S_1, and $\{s_i\}$ is called F_2. The functions are denoted as $F_1 \rightarrow F_2$ and $F_1 \rightarrow S_1$.

We note that $f_{F_1}(j,i,M_t)$ denotes the proper subset of F, and similarly in the sequel.

Now, if there is at least one nonempty F_1 in M_t, then there is a largest of these, which we denote by F_1^*. The largest S_1 and F_2 are understood similarly: S_1^* and F_2^*. The functions of M_t are thus $F_1^* \rightarrow F_2^*$, $F_1^* \rightarrow S_1^*$.

System

So far we have mentioned only individual components (endowed with functions) and their sets and have not discussed their relationship to the system. Clearly, if we should write $f_{s_l}(j,i,M_t)$ instead of some $f(j,i,M_t)$, this results in a change of F. For instance, with the appearance of the first functional component, an F^1 replaces F; and, in general, we have a *history* of the system, the sequence of F, F^1, $F^2, \ldots$.

As a consequence, *the system itself is changed* with the appearance of new functions, and there are not just internal changes taking place in the system. Phenomenologically, this is trivial: if both the components and the relations (or transformations) of a system are replaced with others, the identity of the system becomes altered.

More formally, there is a sequence of F mappings in which a new mapping always is determined by the results of the preceding one. This means a recursion. The system then could be identified either with these transformations or with their closure, but it is possible to show here—by considering the mappings as algorithms—that the system's own algorithms are not identical with the one that performs the closure. Thus we cannot commit identity to the latter, and the system's identity changes in the course of iteration (as detailed in Kampis and Csányi 1986b). This is reflected in the definition of *system*: now it is an ordered $O' = (S, F^i, M_t)$ triple, where F^i is not necessarily identical to F. During autogenesis, it is the *system* that changes.

Replicative function, compartment, and AGSP

Consider now those cases where

(a) $$f_{F_1^*}(j,i,M_t) > f(j,i,M_t), \quad s_i \in M_t, \quad s_j \in S_j$$

because we are interested in the active role some components play in the genesis (production) of other components. Eventually we seek a stable functional organization where the functions of components are manifested in the process of producing the components that constitute the system.

First let $\{s_i\} = S_1^*$ and $\{s_j\} = F_2^*$ for the $s_i - s$ and $s_j - s$ of (a); this will show an interesting picture. Let us define sets A_1, A_2, A_3, and A_4 in the following way:

(b) $$A_1 = S_1^* \cap F_2^*$$

(c) $$A_2 = S_1^* \cap F_1^*$$

(d) $$A_3 = F_1^* \cap F_2^*$$

(e) $$A_4 = A_3 \cap S_1^*$$

(remember the meaning of S_1^*, F_1^*, F_2^*).

Now examine the following cases:

(f) $$A_1 \neq \emptyset$$

(g) $$A_2 \neq \emptyset$$

(h) $$A_3 \neq \emptyset$$

(i) $$A_4 \neq \emptyset$$

Formula (f) means that the function of F_1^* increases the probability of genesis of A_1 from A_1; (g) says that the probability of genesis of F_2^* from A_2 is increased due to functions of A_2; according to (h) the probability of genesis of A_3 from S_1^* is increased due to functions of A_3; and (i) says that the functions of A_4 increase the probability of genesis of A_4 from A_4.

Definition six

Functions in formulas (h) and (i) are replicative. This is denoted by $A \xrightarrow{r} A$.

Formula (i) describes the temporal reproduction of a materially closed system. This is called replication in time. Spatial replication occurs where the component set that is endowed with the respective functions is reproduced, and, beyond this, a similar one is produced from external building blocks. Formula (h) leaves open the question of these kinds of replication: it can be any of them, depending on the fate of A_3 (e.g., if it is destroyed in the process, it is temporal replication).

Let A^* denote all of the replicative components of M_t. Although the above A_3 and A_4 are sets of replicative components, it is *not* true that $A \xrightarrow{r} A$ for any $A \subset A^*$: in general, $A \rightarrow A'$, $A' \neq A$.

However, it is possible that for $A'' = A \cup s_l$; $s_l \in A^* - A$ the relation $A'' \xrightarrow{r} A''$ already holds. The sets having this property $A \xrightarrow{r} A$ are in some sense compact (i.e., with respect to replication).

Definition seven

The set $A_k \subset A^*$, $A_k \xrightarrow{r} A_k$ is called a replicative compartment. Replication, according to the present definition, is a logical relationship between objects and the processes producing these objects. This confronts the more conventional view that sees replication as characterized by differential equations of a certain "autocatalytic" type. The basis of the latter view is that reality is "well ordered": immediate causes are sufficient for the characterization of systems. The central dogma of system theory is that the immediate cause of the behavior and modification of objects is the presence or given state of other objects, and that this can be actually expressed in descriptions. These descriptions, however, often do not answer *whys* but only *hows*. The kind of replication, which means positive constants in the main diagonal of the matrix of some first-order ordinary differential equation, is considered by us as *trivial*, and we focus on the logical aspect instead of this kinetical one (for a discussion, see Kampis and Csányi 1986b).

Going back to functions of the type $A \xrightarrow{r} A$, it is important to note that a precise, exact deterministic replication is only a special case. Replication with probability one is therefore called identical replication to distinguish it from the more general form.

With the aid of the concepts of replication and replicative compartments, we are now ready to interpret the Auto Genetic System Precursor (AGSP), which is the smallest A^* that appears in the sequence of systems.

Information

Omitting here the questions of the informational characterization of autogenesis (see Kampis 1986a, 1986b), we mention only that information concepts are tools for the rapid, qualitative characterization of systems; they express the observer's *knowledge* about the system in some condensed form. However, information can *act* within systems at the same time (a clear example is the existence of information-transferring and processing systems, the behavior of which is influenced by information).

Information generally is conceived through a pattern of system states or of state changes. In our analysis, however, it is impossible to express the information content of functional systems in this way, although in these systems there is a manifestation of "something" that should be called information in order to preserve any meaning for the word (Kampis 1986b).

We find information not in the state patterns of systems but in their *transformations*; more exactly we find it in the interrelatedness of transformations and components.

One should distinguish here information-in-the-system from information-of-the-observer (see Kampis 1986b). The first was called *referential* and the latter *nonreferential* information (figure 24). Referential information (Inf1) has a reference in and for the system, and nonreferential information (Inf2) is essentially the concept that is usually called information content. It must be noted that we can

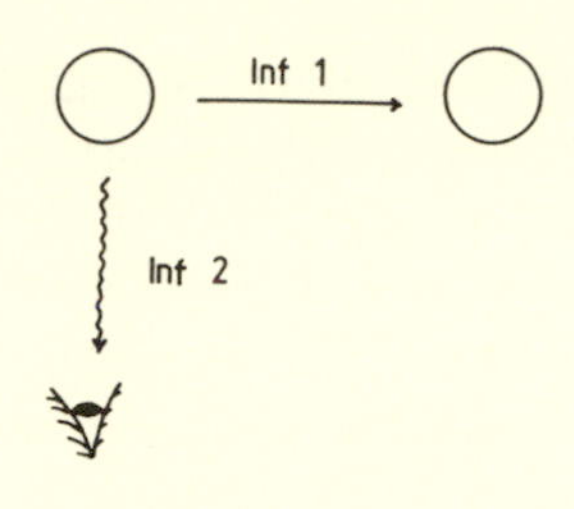

Figure 24

speak of the former only through the latter (for details, see Kampis 1986a, 1986b).

More precisely, it is Infl of figure 24 that so far has been called *function*. Now by the informational characterization of the system we mean the information *content* (Inf2) related to functions, expressed according to some measures. From now on we shall speak about this information.

The information measures presented below serve as aids to a natural interpretation of autogenesis. To this end, unlike in the main text of this book, we shall not concern ourselves with structural descriptions of components here since we can speak about purely transformational descriptions in simpler terms.

When examining the effects of functions we took the zero-system as a base. It is therefore convenient to think that the zero-system also bears some information, determined by the fixed parameters of the system—by S and F—and by M_1. This is called parametric information.

Definition eight

Let M_t be the system state. The parametric information that belongs to component $s_i \in M_t$ is

$$I_P^i = \sum_j f(j,i,M_t) \log f(j,i,M_t) \quad \text{where } s_j \in S_i$$

and the parametric information belonging to M_t is

$$I_P = \sum_i \sum_j f(j,i,M_t) \log f(j,i,M_t); \quad s_i \in M_t, \quad s_j \in S.$$

Summation is understood for those i and j indices, to which F_1^* does not apply. Here and later log means a logarithm on any base.

In the zero-system, F_1^* is empty, and thus the information in fact characterizes the initial system. In the following we shall use I_P as determined at a general time t, i.e., already taking F_1^* into account.

Let us agree that $f'_{F_1}(j,i,M_t)$ should mean the following:

$$f'_{F_1}(j,i,M_t) = \begin{cases} f_{F_1}(j,i,M_t) & \text{if } F_1 \text{ is nonempty} \\ 0 & \text{if } F_1 \text{ is empty.} \end{cases}$$

Functional and replicative information is defined in the following way.

Definition nine

The functional information content of component $s_l \in F_1^*$ in M_t is

$$\begin{aligned} I_f^l = & \sum_i \sum_j f'_{s_l}(j,i,M_t) \log f'_{s_l}(j,i,M_t) \\ & + \sum_P \sum_r f(r,P,M_t) \log f(r,P,M_t); \quad s_i \in M_t, \quad s_j \in S. \end{aligned}$$

Summation is understood for those P-s and r-s which do not appear in f_{s_l}. Functional information belonging to M_t is

$$\begin{aligned} I_f = & \sum_i \sum_j f'_{F_1^*}(j,i,M_t) \log f'_{F_1^*}(j,i,M_t) \\ & + \sum_P \sum_r f(r,P,M_t) \log f(r,P,M_t); \quad s_i \in M_t, \quad s_j \in S \end{aligned}$$

(P and r as above).

We can express the quantity of information belonging to replicative functions in a similar way.

Definition ten

The replicative information content of component $s_l \in F_1^*$ in M_t is

$$\begin{aligned} I_r^l = & \sum_i f'_{s_l}(l,i,M_t) \log f'_{s_l}(l,i,M_t) \\ & + \sum_P f(l,P,M_t) \log f(l,P,M_t); \quad s_i \in M_t. \end{aligned}$$

The replicative information of M_t is

$$I_r = \sum_i f'_{A_3}(A_3, i, M_t) \log f'_{A_3}(A_3, i, M_t) + \sum_P f(A_3, i, M_t) \log f(A_3, i, M_t); \quad s_i \in M_t,$$

(P is as in definition two), and

$$f(A_3, i, M_t) = \sum_{j \in A_3} f(j, i, M_t).$$

To sum up verbally, in functional and replicative information, the probabilities of local transformations are considered as a "pattern," the average information content of which is determined by the classical Shannon formula. The expression for this information contains both the functional and nonfunctional production of the respective components and component sets.

Definitions two and three express that selection or choice is a property of function and replication, and it characterizes this selection. In the zero-system where F_1^* is empty, both I_f and I_r are zero, as can be checked immediately. Considering I_P, the parametric information, as the initial information of the history (autogenesis) of the system, the information *generated* in this process is expressible as $\Delta I = I_f - I_p$. In the zero-system $\Delta I = 0$, but in a system where $F_1^* = M_t$ we can see that $\Delta I = I_f$!

I_r is maximal if $A_3 = F_1^* = M_t$ and

$$\sum_{i \in M_t} f_{M_t}(M_t, i, M_t) = 1;$$

$$\sum_{i \in M_t} f_{M_t}(j, i, M_t) = 0$$

for all $s_j \in M_t$.

It can be seen by substitution that in this case I_f is also maximal, and I_P becomes zero. With this, we have shown the following: In a system that replicates itself identically, (1) parametric information is zero; (2) replicative information content is maximal; (3) functional information content is maximal; and (4) the whole information content of the system is generated *as original*, i.e., it equals the information gain ΔI.

Autogenesis

In autogenesis the processes that take place in the zero-system lead to the development of AGSPs, functional components appear (the open functional network of the system becomes the driving force of further changes); replicative information content and with it functional information content increase; replicative compartments are formed; and eventually the system gets into the state of identical replication described by a closed functional network. Because of this, there is no way for further evolution caused by internal factors. That is:

Definition eleven

Autogenesis is a sequence of systems, or more precisely, of system transformations F, F^1, F^2, ..., F^j, ..., F^s, ..., F^l such that (1) F belongs to the zero-system; (2) M_t, which belongs to F^j, contains an AGSP; (3) M_t, which belongs to F^s, contains replicative compartment(s) different from the AGSP; and (4) M_t, which belongs to F^l, is identically replicative.

Autogenesis comes along with the sequential changes of system identity and with the generation of replicative and functional information. We have devised a computer simulation model of this autogenetic process that can show the phases and phenomena outlined here (Kampis and Csányi 1987b).

References

Alexander, R. O., 1974: "The Evolution of Social Behavior." Ann. Rev. Ecol. Syst. 5:325–383.

———, and G. Borgia, 1978: "Group Selection, Altruism and the Levels of Organization of Life." Ann. Rev. Ecol. Syst. 9:449–474.

Alland, A., 1973: *Evolution and Human Behavior*. 2d ed. Doubleday, New York.

Allen, R., 1982: *How to Save the World?* Corgi Books, London.

Alvarez, L. W., W. Alvarez, F. Asaro, and H. V. Michel, 1980: "Extraterrestrial Cause for the Cretaceous-Tertiary Extinction." Science 208:1095–1108.

Alvarez, W., E. G. Kaufman, F. Surlyk, L. W. Alvarez, F. Asaro, and H. V. Michel, 1984: "Impact Theory of Mass Extinctions and the Invertebrate Fossil Record." Science 208:1135–1141.

Anderson, J. R., 1984: "The Development of Self-Recognition: A Review." Dev. Psychobiol. 17:35–49.

Aoki, K., 1983: "A Quantitative Genetic Model of Reciprocal Altruism." Proc. Natl. Acad. Sci. USA 80:4065–4068.

Arbib, M. A., 1972: *The Metaphorical Brain*. Wiley, New York.

Archer, J., 1976: "The Organization of Aggression and Fear in Vertebrates" in P. P. G. Bateson and P. Klopfer (eds.), *Perspectives in Ethology*, vol. 2. Plenum Press, New York.

Argyll, E., 1977: "Chance and the Origin of Life." Origins of Life 8:287–298.

Ashby, W. R., 1956: *An Introduction to Cybernetics*. Chapman and Hall, London.

Bánáthy, B. H., 1973: *Developing a System View of Education Intersystems*. Intersystems Publications, Seaside, Calif.

Barker, W. C., L. K. Ketcham, and K. O. Dayhoff, 1978: "A Comprehensive Examination of Protein Sequences of Evidence of International Gene Duplication." J. Mol. Evol. 10:265–281.

Barrow, J. D., and F. J. Tipler, 1978: "Eternity Is Unstable." Nature 76:453–459.

Batchinsky, A. G., and V. A. Ratner, 1976: "A Cybernetic Approach to the Origin of the Genetic Coding Mechanism: II. Formation of the Code Series." Origins of Life 7:228–253.

Bauman, P., and E. Patzelt, 1977: *Menschen im Regenwald Expedition Auka*. Fischer Verlag, Frankfurt am Main.

Bawin, S. M., and W. R. Adey, 1976: "Sensitivity of Calcium Binding in Cerebral Tissue to Weak Environmental Electric Fields Oscillating at Low Frequency." Proc. Natl. Acad. Sci. USA 73:1999–2003.

Beadle, G., and M. Beadle, 1966: *The Language of Life*. Doubleday, New York.

Benedict, R., 1934: *Patterns of Culture*. Houghton Mifflin, Boston.

Berger, P. L., and T. Luckmann, 1967: *The Social Construction of Reality*. Anchor Books, New York.

Bergson, H., 1902: *Creative Evolution*. Macmillan, London.

Beritashvili, I. S., 1971: *Vertebrate Memory, Characteristics and Origin*. Plenum Press, New York.

Bertalanffy, L., 1968: *General Systems Theory*. George Braziller, New York.

Bloch, D. P., B. MacArthur, R. Widdowson, D. Spector, R. C. Guimares, and J. Smith, 1983: "tRNA-rRNA Sequence Homologies: Evidence for a Common Evolutionary Origin?" J. Mol. Evol. 19:420–428.

Bloom, L., 1976: "Child Language and the Origin of Language" in S. R. Harnad, H. D. Steklis, and J. Lancaster (eds.), *Origins and Evolution of Language and Speech*. Ann. N.Y. Acad. Sci. 280:170–172.

Bökönyi, S., 1978: *Vadakat terelö Juhász*. Magvetö, Budapest.

Bolin, B., 1981: "Interactions of Biogeochemical Cycles." Nature 293:434.

Bonner, J. T., 1980: *The Evolution of Culture in Animals*. Princeton Univ. Press, Princeton, N.J.

Botkin, D. B., 1982: "Can There Be a Theory of Global Ecology?" J. Theor. Biol. 96:95–98.

Brack, A., and L. E. Orgel, 1975: "Beta-structures of Alternating Polypeptides and Their Possible Prebiotic Significance." Nature 256:383–387.

Brandt Comission, 1983: *Common Crisis*. Pan Books, London.

Brewin, N., 1972: "Catalytic Role of RNA in DNA Replication." Nature 236.

Brown, J. S., M. J. Sanderson, and R. E. Michod, 1982: "Evolution of Social Behavior by Reciprocation." J. Theor. Biol. 99:319–339.

Brown, R., 1973: *A First Language: The Early Stages*. Harvard Univ. Press, Cambridge, Mass.

Bullock, T. M., and G. A. Horrige, 1975: *Structure and Function in the Nervous Systems of Invertebrates*. W. H. Freeman, San Francisco.

Bunge, M., 1963: *The Myth of Simplicity*. Prentice-Hall, Englewood Cliffs, N.J.

Burks, A. W., 1970: *Essay on Cellular Automata*. Univ. of Illinois Press, Urbana.

Burns, B. D., 1958: *The Mammalian Cerebral Cortex*. Arnold, London.

Campbell, D. T., 1960: "Blind Variation and Selective Survival as a General Strategy in Knowledge Processes" in M. C. Jovits and S. Cameron (eds.), *Self-Organizing Systems*. Pergamon, New York.

———, 1974: "Evolutionary Epistemology" in P. A. Schilpp (ed.), *The Philosophy of Karl Popper*. Open Court, Lasalle, Ill.

Canuto, V. M., J. S. Levine, T. R. Augustsson, C. L. Imhoff, and M. S. Giampara, 1983:

"The Young Sun and the Atmosphere and Photochemistry of the Early Earth." Nature 305:281–286.

Caplan, A. L. (ed.), 1978: *The Sociobiological Debate*. Harper and Row, New York.

Caplan, I., and C. P. Ordhal, 1978: "Irreversible Gene Repression Model for Control of Development." Science 201:120–130.

Capon, B., 1964: "Neighbouring Groups Participation." Quart. Res. 18:45–111.

Castle, W. E., 1903: "The Laws of Heredity of Galton and Mendel and Some Laws Governing Race Improvement by Selection." Proc. Am. Acad. Arts and Sci. 39:223–243.

Cavalli-Sforza, L. L. and M. W. Feldman, 1978: "Toward a Theory of Cultural Evolution." Interdiscipl. Sci. Review 3:99–107.

———, 1981: *Cultural Transmission and Evolution*. Princeton Univ. Press, Princeton, N.J.

Cavalli-Sforza, L. L., M. W. Feldman, K. H. Chen, and S. M. Dornbush, 1982: "Theory and Observation in Cultural Transmission." Science 218:19–27.

Chaitin, G., 1966: "On the Length of Programs for Computing Finite Binary Sequences." J. Ass. Comp. Mach. 13:547–569.

Chamblis, G., G. R. Craven, J. Davies, K. Davis, L. Kahan, and M. Nomura (eds.), 1980: *Ribosome: Structure, Function and Genetics*. Univ. Park Press, Baltimore.

Chapple, E. D., 1970: *Culture and Biological Man*. Holt, Rinehart and Winston, New York.

Charlesworth, B., R. Lande, and M. Slatkin, 1982: "A Neo-Darwinian Commentary on Macroevolution." Evolution 36:474–498.

Chevalier-Skolnikoff, S., 1976: "The Ontogeny of Primate Intelligence and Its Implications for Communication Potential" in S. R. Harna, H. D. Steklis, and J. Lancaster (eds.), *Origins and Evolution of Language and Speech*. Ann. N.Y. Acad. Sci. 280:1–914.

Chomsky, N., 1968: *Language and Mind*. Harcourt Brace Jovanovich, New York.

Clark, G., 1970: *Aspects of Prehistory*. Univ. California Press, Berkeley.

Cloak, F. T., Jr., 1975: "Is Cultural Ethology Possible?" Human Ecology 3:161–182.

Codd, E. F., 1968: *Cellular Automata*. Academic Press, New York.

Covey, C., S. H. Schneider, and S. L. Thomson, 1984: "Global Atmospheric Effects of Massive Smoke Injections from a Nuclear War." Nature 308:21–25.

Craik, K. J. W., 1943: *The Nature of Explanation*. Cambridge Univ. Press, Cambridge.

Crick, F., 1967: *Of Molecules and Man*. Univ. of Washington Press, Seattle.

———, 1968: "The Origin of the Genetic Code." J. Mol. Biol. 38:367–379.

Crick, F., S. Brenner, A. Klug, and G. Pilczenik, 1976: "A Speculation on the Origin of Protein Synthesis." Origins of Life 7:389–397.

Crick, F., and L. E. Orgel, 1973: "Directed Panspermia." Icarus 19:341–346.

Cronin, J. E., N. T. Boaz, O. B. Stringer, and Y. Rak, 1981: "Tempo and Mode in Hominid Evolution." Nature 292:113–122.

Croze, H., 1970: "Searching Image in Carrion Crows." Z. Tierpsychology 5:1–86.

Csányi, V., 1978: "Az evolució általános elmélete. Fizikai Szemle 28:401–443.

———, 1979: *Az evolució általános elmélete*. Akadémiai Kiadó, Budapest.

———, 1980: "The General Theory of Evolution." Acta Biol. Hung. Acad. Sci. 31:409–434.

———, 1981: "General Theory of Evolution." Soc. Gen. Syst. Res. 6:73–95.

———, 1982: *General Theory of Evolution*. Publ. House Hung. Acad. Sci., Budapest.

———, 1985: "Autogenesis: Evolution of Self-Organizing Systems" in J. P. Aubin, D. Saari, and K. Sigmund (eds.), "Dynamics of Macrosystems." Proceedings, Laxenburg, Austria, 1984. Lecture notes in Economics and Mathematical Systems no. 257, pp. 253–268. Springer, Berlin.

———, 1985a: "Ethological Analysis of Predator Avoidance by the Paradise Fish (*Macropodus opercularis*): I. Recognition and Learning of Predators." Behaviour 92:227–240.

———, 1985b: "Ethological Analysis of Predator Avoidance by the Paradise Fish (*Macropodus opercularis*): II. Key Stimuli in Avoidance Learning." Anim. Learn. Behav. 14:101–109.

———, 1985c: "How Is the Brain Modelling Its Environment? A Case Study of the Paradise Fish" in G. Montalenti and G. Tecce (eds.), "Variability and Behavioral Evolution." Proceedings, Accademia Nazionale dei Lincei, Rome, 1983, Quaderno no. 259, pp. 142–157.

———, 1987a: "The Replicative Model of Evolution: A General Theory." World Future: J. Gen. Evol. 23:31–65.

———, 1987b: "The Replicative Evolutionary Model of Animal and Human Minds." World Future: J. Gen. Evol. 24:174–214.

Csányi, V., and Judit Gervai, 1986: "Behavior-Genetic Analysis of the Paradise Fish (*Macropodus opercularis*): II. Passive Avoidance Conditioning in Recombinant Inbred Strains." Behav. Genet. 16:553–557.

Csányi, V., and Gy. Kampis, 1985: "Autogenesis: The Evolution of Replicative Systems." J. Theor. Biol. 114:303–321.

Csányi, V., P. Tóth, V. Altbacker, A. Dóka, and Judit Gervai, (1985a): "Behavior Elements of the Paradise Fish (*Macropodus opercularis*): I. Regularities of Defensive Behavior." Acta Biol. Hung. 36:93–114.

———, (1985b): "Behavior Elements of the Paradise Fish (*Macropodus opercularis*): II. A Functional Analysis." Acta Biol. Hung. 36:115–130.

Csendes, T., 1984: "A Simulation Study on the Chemoton." Kybernetes 13:79–85.

Dalenoort, G. J., 1982: "In Search of the Conditions for the Genesis of Cell Assemblies: A Study in Self-Organization." J. Soc. Biol. Struct. 5:161–187.

Darlington, P. I., 1972: "Nonmathematical Concepts of Selection, Evolutionary Energy and Levels of Evolution." Proc. Natl. Acad. Sci. USA 69:1239–1243.

Darwin, C. R., 1859: *The Origin of Species*. John Murray, London.

Dawkins, R., 1974: "Some Descriptive and Explanatory Stochastic Models of Decision Making" in D. J. McFarland (ed.), *Motivational Control Systems*. Analysis Academic Press, London.

———, 1976: *The Selfish Gene*. Oxford Univ. Press, New York.

———, 1978: "Replicator Selection and the Extended Phenotype." Z. Tierpsychol. 47:61–76.

Dayhoff, M. O., E. R. Lippincott, and R. V. Eck, 1964: "Thermodynamic Equilibria in Prebiological Atmospheres." Science 140:1461–1464.

Delsemme, A. H., 1984: "The Cometary Connection with Prebiotic Chemistry." Origins of Life 14:51–60.

Denett, D. C., 1983: "Intentional Systems in Cognitive Ethology: The 'Panglossian Paradigm' Defended." Behav. Brain Sci. 6:343–390.

Doolittle, W. F., and C. Sapienza, 1980: "Selfish Genes, the Phenotype Paradigm and Genom Evolution." Nature 284:601–603.

Dover, G. C., and R. B. Flawell (eds.), 1982: *Genom Evolution*. Academic Press, London.

Drake, J. W., 1974: "The Role of Mutation in Bacterial Evolution." Symp. Soc. Gen. Microbiol. 24:41–58.

Dunbar, M. Y., 1972: "The Ecosystem as a Unit of Natural Selection" in E. S. Deevey (ed.), *Growth by Intussusception: Ecological Essays in Honor of G. Evelyn Hutchinson*. Transactions of the Academy 44:114–130. Connecticut Acad. Arts and Sci., New Haven.

Durham, W. H., 1978: "Toward a Coevolutionary Theory of Human Biology and Culture" in A. L. Caplan (ed.), *The Sociobiological Debate*. Harper and Row, New York.

Duvigneaud, P., 1967: *L'ecologie, Science moderne de Synthése* 2. Ecosystémes et biosphére, Brussells.

Eccles, J. C., 1976: "Brain and Free Will" in G. G. Globus, G. Maxwell, and I. Savodnik (eds.), *Consciousness and the Brain*. Plenum Press, New York.

Edelman, G. M., 1977: "Group Degenerate Selection and Phasic Reentrant Signaling: A Theory of Higher Brain Function" in F. O. Schmitt (ed.), *The Neurosciences: Fourth Study Program*. MIT Press, Cambridge, Mass.

Eibl-Eibesfeldt, I., 1970: *Ethology: The Biology of Behavior*. Holt, Rinehart and Winston, New York.

———, 1979: "Humanethology: Concepts and Implications for Sciences of Man." Behav. Brain Sci. 2:1–57.

———, 1982: "Warfare, Man's Indoctrinability and Group Selection." Z. Tierpsychol. 60:177–198.

Eigen, M., 1971: "Self-Organization of Matter and the Evolution of Biological Macromolecules." Naturwissenschaften 58:465–523.

Eigen, M., and P. Schuster, 1977: "The Hypercycle: A Principle of Natural Self-Organization, Part A: Emergence of the Hypercycle. Naturwissenschaften 64:541–565.

———, 1978a: "The Hypercycle: A Principle of Natural Self-Organization, Part B: The Abstract Hypercycle." Naturwissenschaften 65:7–41.

———, 1978b: "The Hypercycle: A Principle of Self-Organization, Part C: The Realistic Hypercycle." Naturwissenschaften 68:282–369.

Eigen, M., and R. Winkler-Oswatitsch, 1981: "Transfer RNA an Early Gene?" Naturwissenschaften 68:282–292.

Eisenberg, J. F., 1981: *The Mammalian Radiation*. Univ. of Chicago Press, Chicago.

Eldredge, N., 1971: "The Allopatric Model and Phylogeny in Paleozoic Invertebrates." Evolution 25:156–167.

Eldredge, N., and S. J. Gould, 1972: "Punctuated Equilibria: An Alternative to Phyletic Gradualism" in T. J. Schopf (ed.), *Models in Paleobiology*. Freeman, Cooper, San Francisco.

Ellul, J., 1965: *The Technological Society*. Jonathan Cape, London.

Epstein, R., C. E. Kirshnit, R. P. Lanza, and L. C. Rubin, 1984: "'Insight' in the Pigeon: Antecedents and Determinants of an Intelligent Performance." Nature 303, 61–62.

Evarts, E. V., 1967: "Representation of Movements and Muscles by Pyramid Tract Neurons of the Precentral Motor Cortex" in M. D. Yahr and D. P. Purpura (eds.), *The Neurophysiological Basis of Normal and Abnormal Motor Activity*. Raven Press, New York.

Falk, D., 1983: "Cerebral Cortices of East African Early Hominids." Science 221:1072–1074.

Ferracin, A., 1981: "A Neutral Theory of Biogenesis." Origins of Life 11:369–385.

Ferris, J. P., H. Yanagawa, and W. J. Hagan, Jr., 1984: "The Prebiotic Chemistry of Nucleotides." Origins of Life 14:99–106.

Fischer, R. A., 1930: *The Genetical Theory of Natural Selection*. Clarendon Press, Oxford.

Fitch, W., 1973: "Aspects of Molecular Evolution." Ann. Rev. Genet. 7:343–380.

Fox, S. W. (ed.), 1965: *The Origin of Prebiological Systems*. Academic Press, New York.

Fox, S. W., 1973: "Origin of the Cell: Experiments and Premises." Naturwissenschaften 60:359–368.

———, 1980: "The Origins of Behavior in Macromolecules and Protocells." Comp. Biochem. Physiol. 67B:423–436.

———, 1984: "Self-Sequencing of Amino Acids and Origins of Polyfunctional Protocells." Origins of Life 14:485–488.

Fox, S. W., and K. Dose, 1972: *Molecular Evolution and the Origin of Life*. W. H. Freeman, San Francisco.

Frisch, K. von, 1967: *The Dance Language and Orientation of Bees*. Harvard Univ. Press, Cambridge, Mass.

Fuller, J. L., and E. C. Simmel (eds.), 1983: *Behavior Genetics: Principles and Applications*. Lawrence Erlbaum, Hillsdale, N.J.

Fuller, J. L., and W. R. Thompson, 1978: *Foundations of Behavior Genetics*. Mosby, St. Louis.

Gallistel, C. R., 1980: *The Organization of Action: A New Synthesis*. Lawrence Erlbaum, Hillsdale, N.J.

Gallup, G. G., 1970: "Chimpanzees, Self-recognition." Science 167:86–87.

Gánti, T., 1975: "Organization of Chemical Reactions into Dividing and Metabolizing Units: The Chemotons." Biosystems 7:15–21.

———, 1977: *A Theory of Biochemical Supersystems and Its Application for Natural and Artificial Biogenesis*. Akadémiai Kiadó, Budapest.

———, 1980: "On the Organizational Basis of Evolution." Acta Biol. Acad. Sci. Hung. 31:449–459.

Garcia, J., F. R. Ervin, and R. Koelling, 1966: "Learning with Prolonged Delay of Reinforcement." Psychonomic Science 5:121–122.

Gardner, R. A., and B. Gardner, 1969: "Teaching Sign Language to a Chimpanzee." Science 165:664–672.

Gehrz, R. D., D. C. Black, and P. M. Solomon, 1984: "The Formation of Stellar Systems from Interstellar Molecular Clouds." Science 224:823–830.

Gervai, Judit, and V. Csányi, 1985: "Behavior-Genetic Analysis of the Paradise Fish (*Macropodus opercularis*): I. Characterization of the Behavioral Responses of Inbred Strains in Novel Environments: A Factor Analysis." Behav. Genet. 15:503–519.

Ghiselin, M. T., 1981: "Categories, Life and Thinking." Behav. Brain Sci. 4:269–313.

Glaserfeld, E., 1976: "The Development of Language as Purposive Behavior" in S. R. Harnad, H. S. Steklis, and J. Lancaster (eds.), *Origins and Evolution of Language and Speech*. Ann. N.Y. Acad. Sci. 280:212–226.

Glassman, R. B., 1977: "How Can So Little Brain Hold So Much Knowledge?" Psychol. Rec. 1977/2, 393–415.

Goldschmidt, R., 1940: *The Material Basis of Evolution*. Yale Univ. Press, New Haven, Conn.

Goodall, J., and H. van Lawick, 1970: *Innocent Killers*. Ballantine Books, New York.

Goodall-Van Lawick, J., H. Van Lawick, and C. Packer, 1973: "Tool-use in Free-living Baboons in the Gombe National Park, Tanzania." Nature 241:212–213.

Gould, J. L., 1974: "Genetics and Molecular Ethology." Z. Tierpsychol. 36:267–292.

Gould, J. L., and C. G. Gould, 1982: "The Insect Mind: Physics or Metaphysics?" in D. R. Griffin (ed.), *Animal Mind—Human Mind*. Springer, Berlin.

Gould, S. J., 1980: "Is a New and General Theory of Evolution Emerging?" Paleobiology 6:119–130.

Gould, S. J., and N. Eldredge, 1977: "Punctuated Equilibria: The Tempo and Mode of Evolution Reconsidered." Paleobiology 3:115–151.

Gowlet, J. A., J. W. K. Harris, D. Walton, and B. A. Wood, 1981: "Early Archeological Sites: Hominid Remains and Traces of Fire from Chesowania, Kenya." Nature 294:125–129.

Grastyán, E., E. R. John, and I. Bartlett, 1978: "Evoked Response Correlate of Symbol and Significance." Science 201:168–171.

Gray, W., 1975: "Emotional Cognitive Structure Theory and the Development of a General Systems Psychotherapy." General Systems 20:17–23.

Greenberg, J. M., 1984: "Chemical Evolution in Space." Origins of Life 14:25–36.

Griffin, D. R., 1976: *The Question of Animal Awareness*. Rockefeller Univ. Press, New York.

———, 1984: *Animal Thinking*. Harvard Univ. Press, Cambridge, Mass.

Groth, W., 1960: "Photochemical Production of Some Organic Compounds Under Possible Primitive Earth Conditions" in F. Daniels (ed.), *Photochemistry in Liquid and Solid States*. Wiley, New York.

Haldane, J. B. S., 1924: "A Mathematical Theory of Natural and Artificial Selection. Part I." Trans. Camb. Phil. Soc. 23:19–41.

Hamilton, W. D., 1964: "The Genetical Evolution of Social Behavior." J. Theor. Biol. 7:1–52.

Hardy, G. H., 1908: "Mendelian Proportions in a Mixed Population." Science 28:49–50.

Harris, M., 1977: *Cannibals and Kings: The Origins of Cultures*. Random House, New York.

Haukioja, E., 1982: "Are Individuals Really Subordinated to Genes? A Theory of Living Entities." J. Theor. Biol. 99:357–375.

Hebb, D. O., 1946: "On the Nature of Fear." Psychol. Rev. 53:259–276.

Herskowitz, I. H., 1973: *Principles of Genetics*. Macmillan, New York.

Hewes, G. W., 1971: "An Explicit Formulation of the Relationship Between Tool-using, Tool-making and the Emergence of Language" in Abstracts, Am. Anthrop. Assoc. Ann. Meetings, New York.

Hilgard, E. R., and Marguis, D. G., 1940: *Conditioning and Learning*. Appleton, New York.

Hinde, R. A., 1983: *Primate Social Relationships*. Blackwell, Oxford.

Hinde, R. A., and J. Fischer, 1952: "Further Observations on the Opening of Milk Bottles by Birds." Brit. Birds 44:306–311.

Hinde, R. A., and J. Stevenson-Hinde, 1973: *Constraints on Learning*. Academic Press, New York.

Ho, M., and P. T. Saunders, 1979: "Beyond Neo-Darwinism: An Epigenetic Approach to Evolution." J. Theor. Biol. 78:573–591.

Hockett, C. F., 1960: "Logical Considerations in the Study of Animal Communications" in W. E. Lanyon and W. N. Tavolga (eds.), *Animal Sounds and Communication*. Amer. Inst. Biol. Sci., Washington, D.C.

Holling, C. S., 1976: "Resilience and Stability of Ecosystems" in E. Jantsch and C. H. Waddington (eds.), *Evolution and Consciousness*. Addison-Wesley, London.

Holmquist, R., and H. Moise, 1975: "Compositional Nonrandomness: A Quantitatively Conserved Evolutionary Invariant." J. Mol. Evol. 6:1–14.

Hostetler, J. A., 1980: *Amish Society*. 3d ed. Johns Hopkins Univ. Press, Baltimore.

Houk, G., and J. J. Geibel, 1974: "Observation of Underwater Tool Use by the Sea Otter." Calif. Fish and Game 60:207–208.

Hoyle, G., and M. Burrolus, 1973: "Neural Mechanisms Underlying Behavior in the Locust (*Schistocerca gregaris*): III. Topography of Limb Motoneurons in the Metathoracic Ganglion." J. Neurobiol. 4:167–168.

Hubel, D. H., and T. N. Wiesel, 1974: "Sequence Regularity and Geometry of Orientation Columns in the Monkey Striate Cortex." J. Comp. Neurol. 158:267–294.

Humphrey, N. K., 1976: "The Social Function of Intellect" in P. P. G. Bateson and R. A. Hinde (eds.), *Growing Points in Ethology*. Cambridge Univ. Press, Cambridge.

Hutchinson, G. E., 1954: "The Biochemistry of the Terrestrial Atmosphere" in G. Kupik (ed.), *The Earth as a Planet*. Univ. of Chicago Press, Chicago.

———, 1978: *An Introduction to Population Ecology*. Yale Univ. Press, New Haven, Conn.

Iberall, A. S., 1983: "What Is 'Language' That Can Facilitate the Flow of Information? A Contribution to a Fundamental Theory of Language and Communication." J. Theor. Biol. 102:347–359.

Inoue, T., and L. E. Orgel, 1983: "A Nonenzymatic RNA Polimerase Model." Science 219:859–862.

Irvine, W. M., and A. Hjalmarson, 1984: "The Cultural Composition of Interstellar Molecular Clouds." Origins of Life 14:15–23.

Isaac, G. L., 1976: "Stages of Cultural Elaboration in the Pleistocene: Possible Archeological Indicators of the Development of Language Capabilities" in S. R. Harnad, D. H. Steklis, and J. Lancaster (eds.), *Origins and Evolution of Language and Speech*. Ann. N.Y. Acad. Sci. 280:131–142.

———, 1978: "The Food-sharing Behavior of Protohuman Hominids." Sci. Am. 238: 90–110.

Ishigami, M., O. Aonu, T. Hamamoto, M. Kinjo, S. Saigo, K. Gotto, and Y. Hattori, 1984: "The Selection and Coexistence of a Plural Number of Primitive TRNAS and the Origin of the Genetic Code." Origins of Life 14:605–612.

IUCN, 1975: *Red Data Book*. IUCN, Gland.

———, 1978: *The* IUCN *Plant Red Data Book*. IUCN, Gland.

Jantsch, E., and C. H. Waddington, 1976: *Evolution and Consciousness: Human Systems in Transition*. Addison-Wesley, London.

Jenkins, P. F., 1977: "Cultural Transmission of Song Patterns and Dialect Development in a Free-living Bird Population." Anim. Behav. 25:50–78.

Jerison, H. J., 1973: *Evolution of the Brain and Intelligence*. Academic Press, New York.

———, 1978: "The Evolution of Consciousness." The Seventh Int. Conf., Unity of Sci., Boston.

John, E. R., 1967: *Mechanism of Memory*. Academic Press, New York.

———, 1972: "Switchboard versus Statistical Theories of Learning and Memory." Science 172:850–864.

John, E. R., et al., 1977: "Neurometrics." Science 196:1393–1410.

Johnson-Laird, P. N., C. Robins, and L. Velicogna, 1974: "Memory for Words." Nature 251:704–709.

Jukes, T. H., 1967: "Indications of an Evolutionary Pathway in the Amino Acid Code." Biochem. Biophys. Res. Comm. 27:573–578.

Kampis, Gy., 1986a: "Biological Information as a System Description" in R. Trappl (ed.), *Cybernetics and Systems '86*. D. Reidel, Dordrecht.

———, 1986b: "Problems of System Descriptions: I. Function." Int. J. Gen. Syst. 13:143–146.

———, 1986c: "Problems of Systems Descriptions: II. Information." Int. J. Gen. Syst. 13:157–171.

Kampis, Gy., and V. Csányi, 1985: "Simple Models Don't Eliminate Complexity from the Real World." J. Theor. Biol. 115:467–469.

———, 1987a: "Notes on Order and Complexity." J. Theor. Biol. 124:111–121.

———, 1987b: "A Computer Model of Autogenesis." Kybernetes 16:169–181.

———, 1987c: "Replication in Abstract and Natural Systems." Biosystems 20:143–156.

Kandel, E., 1976: *The Cellular Basis of Behavior*. W. H. Freeman, San Francisco.

Katchalsky, A., and P. F. Curran, 1965: *Nonequilibrium Thermodynamics*. North-Holland, Amsterdam.

Kawai, M., 1965: "Newly Acquired Precultural Behavior of the Natural Troops of Japanese Monkeys on Koshima Island." Primates 6:1–30.

Kawamura, S., 1963: "The Processes of Subcultural Propagation Among Japanese Macaques" in I. Southwick (ed.), *Primate Social Behavior*. Van Nostrand, New York.

Kimborough, S. O., 1980: "The Concepts of Fitness and Selection in Evolutionary Biology." J. Biol. Soc. Struct. 3:149–170.

Kimura, M., 1968: "Evolutionary Rate at the Molecular Level." Nature 217:624–626.

———, 1969: "The Rate of Molecular Evolution Considered from the Standpoint of Population Genetics." Proc. Natl. Acad. Sci. USA 63:1181–1188.

———, 1976: "How Genes Evolve: A Population Geneticist's View." Ann. Genet. 19:153–168.

———, 1977: "Preponderance of Synonymous Changes as Evidence for the Neutral Theory of Molecular Evolution." Nature 267:275–276.

Kimura, M., and Ohta, T., 1971: "On the Rate of Molecular Evolution." J. Mol. Evol. 1:1–17.

King, G. A. M., 1977: "Symbiosis and the Origin of Life." Origins of Life 8:39–53.

———, 1981: "Growth of a Hypercycle and Comparison with Conventional Autocatalysis." Biosystems 13:225–234.

———, 1982: "Recycling, Reproduction and Life Origins." Biosystems 15:87–89.

King, J. L., and T. H. Jukes, 1969: "Non-Darwinian Evolution." Science 164:788–798.

King, M. C., and A. C. Wilson, 1975: "Evolution of Two Levels in Humans and Chimpanzees." Science 188:107–116.

King's College Sociobiological Group (ed.), 1982: *Current Problems in Sociobiology*. Cambridge Univ. Press, Cambridge.

Koch, A. L., 1984: "Evolution as the Number of Gene Copies Per Primitive Cell." J. Mol. Evol. 20:71–76.

Kolmogorov, A. N., 1965: "Three Approaches for Defining the Concept of Information Quantity." Problems of Information Transmission 1:3–7.

Krechevsky, I., 1932: "Hypotheses in Rats." Psychological Review 39:516–532.

Kropotkin, P., 1902/1972: *Mutual Aid: A Factor of Evolution*. NYU Press, New York.

Kruijt, J. P., 1964: "Ontogeny of Social Behavior in Burmese Red Jungle Fowl." Behavior Suppl. 12:1–201.

Kuhn, H., 1976: "Model Considerations of Life." Naturwissenschaften 63:68–80.

Kupier, T. B. H., and M. Morris, 1977: "Searching for Extraterrestrial Civilizations." Science 196:616–620.

Lacey, J. C., and D. W. Mullins, Jr., 1983: "Experimental Studies Related to the Origin of the Genetic Code and the Process of Protein Synthesis: A Review." Origins of Life 13:1–42.

Laing, R., 1977: "Automaton Models of Reproduction by Self-Inspection." J. Theor. Biol. 66:437–456.

Lajtha, A., 1974: "Amino Acid Transport in the Brain in Vivo and in Vitro in Aromatic Amino Acids in the Brain." Ciba Found. Symp. 22, ASP Elsevier, Excerpta Medica, North-Holland, Amsterdam.

Lakatos, I., 1976: *Proofs and Refutations: The Logic of Mathematical Discovery*. Cambridge Univ. Press, Cambridge.

Lake, J. A., E. Henderson, M. Oakes, and M. W. Clark, 1984: "Eocytes: A New Ribosome Structure Indicates a Kingdom with a Close Relationship to Eucariotes." Proc. Natl. Acad. Sci. USA 81:3786–3790.

Lashley, K., 1950: "In Search of the Engram" in *Physiological Mechanism in Animal Behavior*. Soc. of Exptl. Biol. Symp. no. 4, Cambridge.

Laszlo, E., 1969: *System, Structure and Experience: Toward a Scientific Theory of Mind*. Gordon and Breach, New York.

———, 1972: *Introduction to Systems Philosophy*. Gordon and Breach, New York.

———, (ed.), 1973: *The World System: Models, Norms, Applications*. George Braziller, New York.

———, 1978: *The Inner Limits of Mankind*. Pergamon, Oxford.

———, 1986: *Evolution*. Pergamon, New York.

Lee, R., 1969: "!Kung Bushmen Subsistence: An Input-Output Analysis" in P. Vayda (ed.), *Environment and Cultural Behavior*. Natural History Press, Garden City, N.Y.

Lehr, E., 1967: "Experimentelle Untersuchungen an Affen und Halbaften über generalisation von insekten und blüttenabbildungen." Z. Tierpsychol. 24:208–244.

Lesk, A. M., 1970: "On the Possibility of a Stage in the Evolution of the Genetic Message in Which Replication Was Imprecise." Biochem. Biophys. Res. Com. 38:855–858.

Lévi-Strauss, C., 1963: *Structural Anthropology*. Basic Books, New York.

Levinton, J. S., 1979: "A Theory of Diversity Equilibrium and Morphological Evolution." Science 204:335–336.

Lewontin, R. C., 1979: "Sociobiology as an Adaptationist Program." Behav. Sci. 24:5–14.

Liben, L. S., 1976: "Memory from a Cogitive-Developmental Perspective" in W. F. Overton and J. M. Callanger (eds.), *Knowledge and Development*, vol. 1. Plenum Press, New York.

Liberman, E. A., 1979: "Analog-Digital Molecular Cell Computer." Biosystems 11:111–134.

Lieberman, P., 1972: *The Speech of Primates*. Mouton, the Hague.

Lorenz, K., 1965: *Evolution and Modification of Behavior*. Univ. of Chicago Press, Chicago.

———, 1973: *Die Rückseite des Spiegels*. R. Pieper, Munich.

———, 1975: "Kant's Doctrine of the A Priori in the Light of Contemporary Biology" in R. I. Evans (ed.), *Konrad Lorenz: The Man and His Ideas*. Harcourt Brace Jovanovich, New York.

———, 1981: *The Foundations of Ethology*. Springer, New York.

Lotka, A. J., 1925: *Elements of Physical Biology*. Williams and Wilkins, Baltimore.

Luhman, N., 1982: "The World Society as a Social System." Int. J. Gen. Syst. 8:131–138.

Lumsden, C. J., and E. O. Wilson, 1981: *Genes, Mind and Culture*. Harvard Univ. Press, Cambridge, Mass.

Lynch, A., 1983: "Abstract Evolution" (unpublished manuscript).

MacArthur, R., 1971: "Fluctuations of Animal Populations and a Measure of Community Stability." Ecology 36:533–536.

MacArthur, R., and E. O. Wilson, 1963: "An Equilibrium Theory of Insular Zoogeography." Evolution 17:373–387.

MacGrew, W. C., and C. E. G. Tutin, 1973: "Chimpanzee Tool Use in Dental Grooming." Nature 241:477–478.

MacKay, D. M., 1951–52: "Mindlike Behavior of Artifacts." Brit. J. Phil. Sci. 2:105–121.

———, 1965: "A Mind's Eye View of the Brain." Prog. Brain Res. 17:321–332.

———, 1966: "Cerebral Organization and the Conscious Control of Action" in J. Eccles (ed.), *Brain and Conscious Experience*. Springer-Verlag, New York.

MacNeish, R., 1972: *The Prehistory of Tehucan Valley*, vol. 4. Univ. of Texas Press, Austin.

Malinowski, B., 1944: *A Scientific Theory of Culture and Other Essays*. Univ. of North Carolina Press, Chapel Hill.

Margalef, F., 1963: "On Certain Unifying Principles in Ecology." Am. Nat. 97:357–374.

———, 1972: "Homage to Evelyn Hutchinson, or Why There Is an Upper Limit to Diversity?" in E. S. Deevey (ed.), *Growth by Intussusception: Ecological Essays in Honor of G. Evelyn Hutchinson*. Transactions of the Academy 44:213–345. Connecticut Academy of Arts and Sci., New Haven.

Margulis, L., 1970: *Origin of Eucariotic Cells*. Yale Univ. Press, New Haven, Conn.

Margulis, L., and J. L. Lovelock, 1974: "Biological Modulation of the Earth's Atmosphere." Icarus 21:471–489.

Markert, C. L., J. B. Shaklee, and G. S. Whitt, 1975: "Evolution of a Gene." Science 189:102–114.

Marshak, A., 1976: "Some Implications of the Paleolithic Symbolic Evidence for the Origin of Language" in S. R. Harnad, H. D. Steklis, and J. Lancaster (eds.), *Origins and Evolution of Language and Speech*. Ann. N.Y. Acad. Sci. 280:289–311.

Martell, A. E., 1968: "Catalytic Effects of Metal Chelate Compounds." Pure Appl. Chem. 17:129–178.

Matsuno, K., 1974: "Probabalistic Kinetics of Macroprocesses in Broken Microscopic Reversibility." J. Stat. Physics 11:87–132.

———, 1977: "Prebiological Molecular Evolution: A Physicochemical Process." J. Phys. Soc. Jap. 42:1691–1693.

———, 1978: "Appearance of Looped Reactions in Dissipative Chemical Kinetics." Physics Letters 68A: 490–491.

———, 1980: "Compartmentalization of Self-Reproducing Machines: Multiplication of Microsystems with Self-Instructing Polimerization of Amino Acids." Origins of Life 10:361–370.

———, 1981: "Material Self Assembly as a Psychochemical Process." Biosystems 13: 237–241.

———, 1983: "Evolutionary Changes in the Information Content of Polypeptides." J. Theor. Biol. 105:185–199.

———, 1984: "Protobiology: A Theoretical Synthesis" in K. Matsuno et al. (eds.), *Molecular Evolution and Protobiology*. Plenum Press, New York.

Maturana, H. R., and F. J. Varela, 1980: *Autopoiesis and Cognition*. D. Reidel, London.

Maynard-Smith, J., 1964: "Group Selection and Kin Selection." Nature 201:1145–1147.

———, 1969: *The Theory of Evolution*. Penguin, London.

———, 1978: *The Evolution of Sex*. Cambridge Univ. Press, Cambridge.

Mayr, E., 1963: *Animal Species and Evolution*. Harvard Univ. Press, Cambridge, Mass.

———, 1982: *The Growth of Biological Thought*. Harvard Univ. Press, Cambridge, Mass.

Meadows, D., J. Richardson, and G. Bruckman, 1982: *Groping in the Dark*. Wiley, New York.

Mellen, S. W. 1981: *The Evolution of Love*. W. H. Freeman, San Francisco.

Menzel, E. W., and M. K. Johnson, 1976: "Communication and Cognitive Organization in Human and Other Animals" in S. R. Harnad, H. D. Steklis, and J. Lancaster (eds.), *Origins and Evolution of Language and Speech*. Ann. N.Y. Acad. Sci. 280:131–142.

Mesarovic, M. D. (ed.), 1964: *Views on General Systems Theory*. Wiley, New York.

Michod, R. E., 1981: "Positive Heuristics in Evolutionary Biology." Brit. J. Phil. Sci. 32:1–36.

Milgram, S., 1963: "Behavioral Study of Obedience." J. Abnormal Soc. Psychol. 67:371–378.

———, 1965: "Some Conditions of Obedience and Disobedience to Authority." Human Relations 18:57–76.

Miller, G. A., E. Galanter, and K. H. Pribram, 1960: *Plans and Structure of Behavior*. Holt, Rinehart and Winston, New York.

Milner, P. M., 1976: "Models of Motivation and Reinforcement" in A. Wanguing and E. T. Rolls (eds.), *Brain Stimulation Reward*. Elsevier, New York.

———, 1977: "Theories of Reinforcement, Drive, and Motivation" in L. Iversen, S. Iversen, and S. H. Synder (eds.), *Handbook of Psychopharmacology*, vol. 7. Plenum Press, New York.

Mizutani, H., and C. Ponnamperuma, 1977: "The Evolution of the Protein Synthesis System." Origins of Life 8:183–319.

Mizutani, H., and E. Wada, 1982: "Material Cycles and Organic Evolution." Origins of Life 12:369–376.

Monod, J., 1971: *Chance and Necessity*. Knopf, New York.

Montagu, A., 1976: "Toolmaking, Hunting and the Origin of Language and Speech" in S. R. Harnad, H. D. Steklis, and J. Lancaster (eds.), *Origins and Evolution of Language and Speech*. Ann. N.Y. Acad. Sci. 280:226–274.

Morgan, C. L., 1896: *Habit and Instinct*. Edward Arnold, London.

Morowitz, H. J., 1968: *Energy Flow in Biology*. Academic Press, New York.

———, 1971: "An Energetic Approach to Prebiological Chemistry" in R. Buvet and C. Ponnamperuma (eds.), *Chemical Evolution and the Origin of Life*. North-Holland, Amsterdam.

Morris, D., 1962: *The Biology of Art*. Knopf, London.

Mountcastle, V. M., 1967: "The Problem of Sensing and the Neural Coding of Sensory Events" in G. C. Quarton, T. Melnechuk, and F. O. Smitt (eds.), *The Neurosciences: A Study Program*. Rockefeller Univ. Press, New York.

Müller, G. E., and A. Pilzecker, 1900: "Experimentelle Beitrage zur Lehre vom Gedachtnis." Z. Psychol. Suppl. 1:1–288.

Müller-Herold, U., 1983: "What Is a Hypercycle?" J. Theor. Biol. 102:569–584.

Mumford, L., 1967: *The Myth of the Machine*. Harcourt Brace Jovanovich, New York.

Mundinger, P. C., 1980: "Animal Cultures and a General Theory of Cultural Evolution." Ethol. Sociobiol. 1:183–223.

Murdock, G. P., 1967: *Atlas of Ethnography*. Univ. of Pittsburgh Press, Pittsburgh.

Murray, R. D., 1980: "The Evolution and Functional Significance of Incest Avoidance." J. Hum. Evol. 9:173–178.

Myers, N., 1979: *The Sinking Ark*. Pergamon, New York.

Nagy, B., 1976: "Organic Chemistry on the Young Earth." Naturwissenschaften 63:499–505.

Nagy, L. N., 1974: "Transvaal Stromatolite: First Evidence for the Diversification of Cells About 2,2- 10-MY-Years Ago." Science 183:514–515.

Neumann, J. von, 1966: *Theory of Self-Reproducing Automata*. Univ. of Illinois Press, Urbana.

Nottebohm, F., 1977: "Asymmetries in Neural Control of Vocalization in the Canary" in S. Harnad (ed.), *Lateralization in the Nervous System*. Academic Press, New York.

Noyes, W. A., and P. A. Leighton, 1941: *The Photochemistry of Gases*. Reinhold, New York.

Odum, E. P., 1969: "The Strategy of Ecosystem Development." Science 164:262–270.

Oehler, D. Z., J. W. Schof, and K. A. Kvenvolden, 1972: "Carbon Isotopic Studies of Organic Matter in Precambrian Rocks." Science 175:1246–1248.

Ohno, S., 1970: *Evolution by Gene Duplication*. Springer-Verlag, New York.

———, 1984: "Birth of a Unique Enzyme from an Alternative Reading Frame of the Preexisted Internally Repetitious Coding Sequence." Proc. Natl. Acad. Sci. USA 81:2421–2425.

O'Keefe, J., and L. Nadel, 1974: "Maps in the Brain." New Scientist 62:749–751.

Olson, C. B., 1981: "A Theory of the Origin of Life." Origins of Life 11 1:353–368.

Olton, D. S., and R. J. Samuelson, 1976: "Remembrance of Places Passed: Spatial Memory in Rats." J. Exp. Psych.: Anim. Behavior Proc. 2:97–116.

Onsager, L., 1931: "Reciprocal Relations in Irreversible Processes II." Phys. Rev. 38: 2265–2279.

Oparin, A., 1924: *Proizhozsdjenyie zsiznyi*. Moszkovszkij Rabocsij, Moscow.

Orgel, L. E., 1968: "Evolution of the Genetic Apparatus." J. Mol. Biol. 38:380–393.

Orgel, L. E., and F. H. C. Crick, 1980: "Selfish DNA: The Ultimate Parasite." Nature 280:604–607.

Oro, J., K. Rewers, and D. Odom, 1982: "Criteria for the Emergence and Evolution of Life in the Solar System." Origins of Life 12:285–305.

Osgood, C., 1959: *Ingalik Mental Culture*. Yale Univ. Publ. Anthropol. no. 56.

Papagiannis, M. D., 1984: "Life-Related Aspects of Stellar Evolution." Origins of Life 14:43–50.

Pask, G., 1972: "Learning Strategies, Memories and Individuals" in W. H. Robinson, and

D. E. Knight (eds.), *Cybernetics, Artificial Intelligence and Ecology*. Proc. Fourth Ann. Symp. Am. Soc. Cyber. Spartan Books, New York–Washington, D.C.

———, 1975: *Conversation, Cognition and Learning*. Elsevier, Amsterdam.

Passingham, R., 1982: *The Human Primate*. W. H. Freeman, San Francisco.

Pattee, H. H., 1965: "The Recognition of Hereditary Order in Primitive Chemical Synthesis" in S. W. Fox (ed.), *The Origin of Prebiological Systems*. Academic Press, New York.

———, 1967: "Quantum Mechanisms, Heredity and the Origin of Life." J. Theor. Biol. 17:410–417.

———, 1973: "Physical Problems of the Origin of Natural Controls" in A. Locker (ed.), *Biogenesis, Evolution, Homeostasis*. Springer-Verlag, Berlin.

———, 1977: "Dynamic and Linguistic Modes of Complex Systems." Int. J. Gen. Syst. 3:259–266.

Patterson, F., 1978: "Conversations with a Gorilla." National Geographic 154:438–465.

Penfield, W., and L. Roberts, 1966: *Speech and Brain-Mechanisms*. Atheneum, New York.

Pflug, H. D., and H. Jaeschke-Boyer, 1979: "Combined Structural and Chemical Analysis of 3.800-MY-Old Microfossils." Nature 280:483–486.

Piaget, J., 1952: *The Origins of Intelligence in Children*. Internat. Univ. Press, New York.

———, 1954: *The Construction of Reality in the Child*. Ballantine Books, New York.

———, 1960: *The Psychology of Intelligence*. Littlefield, New York.

———, 1967: *Biologie et connaissance*. Gallimard, Paris.

Piaget, J., and B. Inhelder, 1973: *Memory and Intelligence*. Basic Books, New York.

Pimentel, D., 1961: "Animal Population Regulation by Genetic Feed-back Mechanisms." Am. Nat. 93:65–79.

Plotkin, H. C., and F. J. Odling-Smee, 1981: "A Multiple-level Model of Evolution and Its Implications for Sociobiology." Behav. Brain Sci. 4:225–268.

Polányi, M., 1968: "Life's Irreducible Structure." Science 160:1308–1312.

Premack, D., 1971: "Language in Chimpanzees?" Science 172:808–822.

———, 1972: "Cognitive Principles." Paper read at Learning Conference, North Carolina State University, Raleigh.

———, 1980: "Representational Capacity and Accessibility of Knowledge: The Case of Chimpanzees" in Piatelli-Palmerini (ed.), *Language and Learning: The Debate Between Jean Piaget and Noam Chomsky*. Harvard Univ. Press, Cambridge, Mass.

Pribram, K. H., 1976: "Problems Concerning the Structure of Consciousness" in G. G. Globus, G. Maxwell, and I. Savodnik (eds.), *Consciousness and the Brain*. Plenum Press, New York.

Pribram, K. H., R. Baron, and M. Nuwer, 1974: "The Holographic Hypothesis of Memory Structure in Brain Function and Perception" in R. C. Atkinson et al. (eds.), *Contemporary Developments in Mathematical Psychology*. W. H. Freeman, San Francisco.

Prigogine, I., 1955: *Introduction in Thermodynamics of Irreversible Processes*. Charles C Thomas, Springfield, Ill.

———, 1976: "Order Through Fluctuation: Self-Organization and Social Systems" in E. Jantsch and C. H. Waddington (eds.), *Evolution and Consciousness*. Addison-Wesley, London.

Prigogine, I., G. Nicolis, and A. Babloyantz, 1972: "Thermodynamics of Evolution." Physics Today 25:23–28, 38–44.

Prigogine, I., and I. Stengers, 1984: *Order Out of Chaos*. Bantam, New York.

Prigogine, J., and J. M. Wiame, 1946: "Biologie et thermodynamique des phenom's irreversible." Experientia 2:451–453.

Primas, H., 1977: "Theory Reduction and Non-Boolean Theories." J. Math. Biol. 4:281–301.

Pringle, J. W. S., 1951: "On the Parallel Between Learning and Evolution." Behaviour 3:174–214.

Quastler, H., 1964: *The Emergence of Biological Organization*. Yale Univ. Press, New Haven, Conn.

Raff, R. A., and H. R. Mahler, 1972: "Non Symbiotic Origin of Mitochondria." Science 177:575–582.

Rakic, P., 1975: "Local Circuit Neurons." N.R.P. Bulletin 13, no. 3.

Ratner, V. A., and A. G. Batchinsky, 1976: "A Cybernetic Approach to the Origin of the Genetic Mechanisms: I. Methodological Principles." Origins of Life 7:225–228.

Reek, G. R., E. Swanson, and D. C. Teller, 1978: "The Evolution of Histones." J. Mol. Evol. 10:309–317.

Reichert, T. A., I. M. Yu, and R. A. Christensen, 1976: "Molecular Evolution as a Process of Message Refinement." J. Mol. Evol. 8:41–54.

Rhodes, F. H. T., 1983: "Gradualism, Punctuated Equilibrium and the Origin of Species." Nature 305:269–272.

Richardson, D., 1976: "Continuous Self-Reproduction." J. Comp. Syst. Sci. 12:6–12.

Richardson, J. (ed.), 1984: *Models of Reality*. Lomond, Mount Airy, Md.

Riedl, R., 1978: *Order in Living Organisms*. Wiley, New York.

Robinson, B. J., 1976: "Molecular Astronomy." Proc. Astron. Soc. Aust. 3:12–19.

Roheting, D. L., and S. W. Fox, 1967: "The Catalytic Activity of Thermal Polyanhydro &-Amino Acids for the Hydrolisis of p-Nitrophenyl Acetate." Arch. Biochem. Biophys. 118:122–143.

Rohles, F. H., and J. V. Devine, 1966: "Chimpanzee Performance on a Problem Involving the Concept of Middleness." Anim. Behav. 14:159–162.

———, 1967: "Further Studies of the Middleness Concept with the Chimpanzee." Anim. Behav. 15:107–112.

Root-Bernstein, R. S., 1983: "Protein Replication by Amino Acid Pairing." J. Theor. Biol. 100:99–106.

Roper, M. K., 1969: "A Survey of Evidence for Infrahuman Killing in the Pleistocene." Current Anthropology 10:427–459.

Rosen, R., 1973: "On the Generation of Metabolic Novelties in Evolution" in A. Locker (ed.), *Biogenesis, Evolution, Homeostasis*. Springer, Berlin.

———, 1977: "Complexity as a System Property." Int. Gen. Syst. 3:227–232.

Russel, W., and O. P. Nathan, 1946: "Traumatic Amnesia." Brain 69:280–300.

Sahal, D., 1976: "System Complexity: Its Conception and Measurement in the Design of Engineering Systems." IEEE Trans. Syst. Man and Cybernet. 440–445.

Sahlins, M., 1976: *The Use and Abuse of Biology*. Univ. of Michigan Press, Ann Arbor.

Salati, E., and P. B. Vose, 1984: "Amazon Basin: A System in Equilibrium." Science 225:129–138.

Saunders, P. T., and M. W. Ho, 1976: "On the Increase in Complexity in Evolution." J. Theor. Biol. 63:375–384.

———, 1981: "On the Increase in Complexity in Evolution: II. The Relativity of Complexity and the Principle of Minimum Increase." J. Theor. Biol. 90:515–530.

Schlesinger, G., and S. L. Miller, 1983a: "Prebiotic Synthesis in Atmospheres Containing CH_4, CO_2 and CO: I. Amino Acids." J. Mol. Evol. 19:376–382.

———, 1983b: "Prebiotic Synthesis in Atmospheres Containing CH_4, CO_2 and CO: II. Hydrogen Cyanide, Formaldehyde and Ammonia." J. Mol. Evol. 19:383–390.

Schmitt, F. O., P. Dev, and B. H. Smith, 1976: "Electronic Processing of Information by Brain Cells." Science 193:114–120.

Schopf, J. W., 1978: "The Evolution of the Earliest Cells." Sc. Am. 239:84–104.

Schopf, T. J., 1982: "A Critical Assessment of Punctuated Equilibria: I. Duration of Taxa." Evolution 36:1144–1157.

Schuster, P., 1984: "Evolution Between Chemistry and Biology." Origins of Life 14:3–14.

Schwarcz, H. P., and I. Skoflek, 1982: "New Dates for Tata Hungary Archeological Site." Nature 295:590–591.

Schwartz, R. M., and M. O. Dayhoff, 1978: "Origins of Procariotes, Eucariotes, Mitochondria and Chloroplast." Science 199:395–403.

Sebeok, T. A., and J. Umiker-Sebeok (eds.), 1980: *Speaking of Apes*. Plenum Press, New York.

Shapiro, R., 1984: "The Improbability of Prebiotic Nucleic Acid Synthesis." Origins of Life 14:565–570.

Sibley, C. G., and J. E. Ahlquist, 1984: "The Phylogeny of the Hominoid Primates, as Indicated by DNA-DNA Hybridization." J. Mol. Evol. 20:2–15.

Simon, A., 1962: "The Organization of Complex Systems." Proc. Am. Phil. Soc. 106: 467–482.

Simpson, G. G., 1944: *Tempo and Mode in Evolution*. Columbia Univ. Press, New York.

———, 1953: *The Major Features of Evolution*. Columbia Univ. Press, New York.

Skinner, B. F., 1966: "The Phylogeny and Ontogeny of Behavior." Science 153:1205–1213.

———, 1968: "Are Theories of Learning Necessary?" in A. C. Catania (ed.), *Contemporary Research in Operant Behavior*. Scott, Foresman, Glenview, Ill.

Smith, T. F., and J. Hertogen, 1980: "An Extraterrestrial Event at the Cretaceous-Tertiary Boundary." Nature 285:198–203.

Smith, T. F., and H. J. Morowitz, 1982: "Between History and Physics." J. Mol. Evol. 18:265–282.

Sokolov, E. N., 1960: "Neuronal Models and the Orienting Reflex" in M. A. B. Brasier (ed.), *The Central Nervous System and Behavior*. Macy Foundation, New York.

Spencer, H., 1862: *First Principles*.

———, 1889: *The Study of Sociology*.

Sperry, R. W., 1976: "Mental Phenomena as Causal Determinants of Brain Function" in G. G. Globus, G. Maxwell, and I. Savodnik (eds.), *Consciousness and the Brain*. Plenum Press, New York.

Spooner, B., and H. S. Mann, 1982: *Desertification and Development: Dryland Ecology in Social Perspective*. Academic Press, New York.

Staddon, J. E. R., and V. L. Simmelhag, 1971: "The 'Superstition' Experiment: Reexamination of Its Implication for the Principles of Adaptive Behavior." Psych. Rev. 78:3–43.

Stahl, W. R., 1965: "Algorithmically Unsolvable Problems for a Cell Automaton." J. Theor. Biol. 8:371–394.

Stanley, S. M., 1975: "A Theory of Evolution Above the Species Level." Proc. Natl. Acad. Sci. USA 72:646–650.

———, 1979: *Macroevolution: Pattern and Process*. W. H. Freeman, San Francisco.

Stebbins, G. L., 1982: "Perspectives in Evolutionary Theory." Evolution 36:1109–1118.

Stent, G. S., 1975: "Limits to the Scientific Understanding of Man." Science 187:1052–1057.

Sturmwasser, F., 1973: quoted in D. Schneider, "Simple Systems: A Complex Challenge to Neurochemistry." Symp. Am. Soc. Neurochem. Fed. Proc. 32:1449–1456.

Stuss, D. T., and W. T. Picton, 1978: "Neurophysiological Correlates of Human Concept Formation." Behav. Biol. 23:135–162.

Szentágothai, J. 1967: "The Anatomy of Complex Integrative Units in the Nervous System" in K. Lissák (ed.), *Recent Developments of Neurobiology in Hungary*, vol. 1. Akadémiai Kiadó, Budapest.

———, 1972: "The Basic Neuron Circuit of the Neocortex" in N. Petsche-Braiser (ed.), *Symposium on Synchronization Mechanisms*. Springer, Vienna.

———, 1975: "The 'Module-concept' in Cerebral Cortex Architecture." Brain Res. 95:475–496.

———, 1978a: "The Neuron Network of Cerebral Cortex: A Functional Interpretation." Proc. R. Soc. London 201:219–248.

———, 1978b: "The Neuronal Machine of the Cerebral Cortex as a Substrate of Psychic Functions." Int. Congr. Phil. Düsseldorf Abst. vol. 1, 611–614.

———, 1978c: "The Local Neuronal Apparatus of the Cerebral Cortex" in P. A. Buser and A. Rougeul-Buser (eds.), *Cerebral Correlates of Conscious Experience*. North-Holland, Amsterdam.

Symons, D., 1979: *The Evolution of Human Sexuality*. Oxford Univ. Press, Oxford.

Taylor, R. B., 1969: *Cultural Ways*. Allyn and Bacon, Boston.

Teleki, G., 1973: *The Predatory Behavior of Wild Chimpanzees*. Bucknell Univ. Press, Lewisburg, Pa.

———, 1974: "Chimpanzee Subsistence Technology: Materials and Skills." J. Hum. Evol. 3:575–594.

———, 1975: "Primate Subsistence Patterns: Collector Predator and Gatherer Hunters." J. Hum. Evol. 4:125–184.

Terzulo, C. A., and P. Viviani, 1978: "The Central Representation of Learned Motor

Patterns" in R. E. Talbot, and D. R. Humphrey (eds.), *Posture and Movement*. Raven Press, New York.

Tiger, L., 1969: *Men in Groups*. Vintage Books, New York.

Tinbergen, N., 1960: "The Natural Control of Insects in Pine Woods." Arch. Neer. Zool. 13:265–379.

Tolman, E. C., 1932: *Purposive Behavior in Animals and Man*. Appleton-Century-Crofts, New York.

Trivers, R. L., 1971: "The Evolution of Reciprocal Altruism." Quart. Rev. Biol. 46:35–57.

———, 1985: *Social Evolution*. Benjamin/Cummings, Menlo Park, Calif.

Trivers, R. L., and H. Hare, 1976: "Haplodiploidy and the Evolution of the Social Insects." Science 191:249–263.

Turco, R. P., O. B. Toon, T. Ackerman, J. B. Pollock, and C. Sagan, 1983: "Nuclear Winter: Global Consequences of Multiple Nuclear Explosions." Science 222:1283–1292.

Van Valen, L., 1973: "A New Evolutionary Law." Evol. Theor. 1:1–30.

Varela, F. G., H. R. Maturana, and R. Urbibe, 1974: "Autopoiesis: The Organization of Living Systems, Its Characterization and a Model." Biosystems 5:187–196.

Vértes, L., 1964: *Tata: Eine Mittelpaléolitische Travertin in Ungarn*. Series Nova 43 Akadémiai Kiadó, Budapest.

Volterra, V., 1931: *Lecons sur la théorie mathématique de la lutte pour la vie*. Gauthier-Villars, Paris.

Vries, H. de, 1901: *Die Mutationstheorye*.

Waal, F. de. 1983: *Chimpanzee Politics: Power and Sex Among Apes*. Unwin, London.

Walker, T. D., and J. W. Valentine, 1984: "Equilibrium Models of Evolutionary Species Diversity and the Number of Empty Niches." Am. Nat. 124:887–899.

Wallace, D. C., and H. I. Morowitz, 1973: "Genom Size and Evolution." Chromosoma 40:121–126.

Walsh, M. M., and D. R. Lowe, 1985: "Filamentous Microfossils from the 3.500-MY-old Onverwact Group Baberton Mountain Land South Africa." Nature 314:530–532.

Ward, B., and R. Dubos, 1972: *Only One Earth*. Pelican Books, London.

Watanabe, M., 1979: "Prefrontal Unit Activity and Delayed Conditional Discrimination" in B. Agranoff and V. Tsuhada (eds.), *Memory and Learning*. Raven Press, New York.

Watson, J. D., 1976: *Molecular Biology of the Gene*. W. A. Benjamin, New York.

Weinberg, W., 1908: "Über de Nachweis der Vererbung beim Menschen." J. ber. Verein f. Vaterl. Naturk. Württemberg 64:368–382.

Welch, R. G., 1977: "On the Role of Organized Multienzyme Systems in Cellular Metabolism: A General Synthesis." Progr. in Biophys. and Mol. Biol. 32:103–191.

White, H. B., 1976: "Coenzymes as Fossils of an Earlier Metabolic State." J. Mol. Evol. 7:101–104.

Wickelgren, W. A., 1972: "Trace Resistance and the Decay of Long-Term Memory." J. Math. Psychol. 9:418–455.

Wicken, J. S., 1978: "Information Transformation in Molecular Evolution." J. Theor. Biol. 72:191–204.

———, 1979: "The Generation of Complexity in Evolution: A Thermodynamic and Information-Theoretical Discussion." J. Theor. Biol. 77:349–365.

Wigner, E. P., 1961: "Probability of the Existence of a Self-Reproducing Unit in the Light of Personal Knowledge" in *Collection of Essays Presented to M. Polányi*. Routledge and Kegan Paul, London.

Williams, G. O., 1966: *Adaptation and Natural Selection: A Critique of Some Current Evolutionary Thought*. Princeton Univ. Press, Princeton, N.J.

Williams, M. B., 1970: "Deducing the Consequence of Evolution: A Mathematical Model." J. Theor. Biol. 29:343–385.

Willows, A. O. D., 1967: "Behavioral Acts Elicited by Stimulation of Single Identifiable Brain Cells." Science 157:570–574.

Wilson, A. C., S. S. Carlson, and T. J. White, 1977: "Biochemical Evolution." Ann. Rev. Biochem. 46:573–639.

Wilson, D. M., 1970: "Neural Operations in Anthropoid Ganglia" in F. O. Smitt (ed.), *The Neurosciences: Second Study Program*. Rockefeller Univ. Press, New York.

Wilson, E. O., 1973: "Group Selection and Its Significance for Ecology." Biosystems 23:631–638.

———, 1975: *Sociobiology: The New Synthesis*. Harvard Univ. Press, Cambridge, Mass.

Woese, C. R., 1967: *The Genetic Code*. Harper and Row, New York.

———, 1971: "Evolution of Macromolecular Complexity." J. Mol. Theor. Biol. 33:29–34.

———, 1981: "Archebacteria." Sc. Am. 244:94–108.

Woolsey, T. A. and H. Van der Loos, 1970: "The Structural Organization of Layer IV in the Somatosensory Region SI of Mouse Cerebral Cortex: The Description of a Cortical Field Composed of Discrete Cytoarchitectonical Units." Brain Res. 17:205–242.

Wright, S., 1931: "Evolution in Mendelian Populations." Genetics 16:97–159.

———, 1982: "Character Change, Speciation and the Higher Taxa." Evolution 36:427–443.

Wyers, E. J., 1976: "Learning and Evolution" in I. Petrinovich and I. L. McGaugh (eds.), *Knowing, Thinking and Believing*. Plenum Press, New York.

Wynne-Edwards, V. C., 1962: *Animal Dispersion in Relation to Social Behaviour*. Oliver and Boyd, Edinburgh.

Yanagawa, H., and F. Egami, 1980: "Formation of Organized Particles, Marigranules and Marisomes from Amino Acids in a Modified Sea Medium." Biosystems 12:147–154.

Yunis, J. J., J. R. Sawyer, and K. Dunham, 1980: "The Striking Resemblance of High-Resolution G-Banded Chromosomes of Man and Chimpanzee." Science 208:1145–1148.

Zaug, A. J., P. J. Grabowski, and Cech, 1983: "Autocatalytic Cyclization of an Excised Intervening Sequence RNA in a Cleavage Ligation Reaction." Nature 301:578–581.

Zeleny, M., 1977: "Self-Organization of Living Systems: A Formal Model of Autopoiesis." Int. Gen. Syst. 4:13–28.

Zipkas, D., and M. Riley, 1975: "Proposal Concerning Mechanisms of Evolution of the Genome of E. coli." Proc. Natl. Acad. Sci. USA 72:1354–1358.

Zotin, A. J., and R. S. Zotina, 1967: "Thermodynamical Aspects of Developmental Biology." J. Theor. Biol. 17:57–75.

Zuckerkandl, E., 1976a: "Evolutionary Processes and Evolutionary Noises at the Molecular Level: I. Functional Density in Proteins." J. Mol. Evol. 7:167–183.

———, 1976b: "Evolutionary Processes and Evolutionary Noises at the Molecular Level: II. A Selectionist Model from Random Fixations in Proteins." J. Mol. Evol. 7:269–311.

Index

About the Author

Vilmos Csányi is professor of ethology and behavior genetics at the Loránd Eötvös University of Budapest, Hungary. He graduated as a research chemist at the Loránd Eötvös University and received his degrees of Ph.D. and D.Sc. in molecular biology. Presently he is the head of the department of behavior genetics.

He has published more than one hundred research papers and nine books. He also is the author of more than a hundred popular articles on various biological subjects. His recent interests have focused on the system-evolutionary approach to animal behavior and the biological bases and origin of human culture.

Library of Congress Cataloging-in-Publication Data
Csányi, Vilmos, 1935–
Evolutionary systems.
Bibliography: p.
Includes index.
1. Evolution. I. Title.
QH366.2.C76 1988 575.01 88-7149
ISBN 0-8223-0836-3